SOME METHODS FOR THE

Statistical Analysis

of samples of

Benthic Invertebrates

by

J. M. ELLIOTT, Ph.D.
Freshwater Biological Association

FRESHWATER BIOLOGICAL ASSOCIATION
SCIENTIFIC PUBLICATION No. 25
1971

FOREWORD

Those who know about these things may perhaps question the need for yet another handbook on statistical methods—but this booklet is not intended for them. It is rather for the non-statistician, written by a zoologist who found it necessary to teach himself enough about these techniques to apply them with confidence to the particular problems he encountered in analysing his data. Since these data have much in common with those collected by other workers on the bottom fauna of streams and lakes, it seemed useful to others to publish, in a concise form and illustrated by practical examples, what Dr Elliott found out for himself. His aim has been to avoid mathematical argument while making clear both the reasons for choosing one technique rather than another in particular circumstances, and also the limitations to be borne in mind when interpreting the results. The text has been carefully checked by expert statisticians.

THE FERRY HOUSE
2 Dec. 1969

H. C. GILSON
Director

Reprinted with minor corrections, 1973.

CONTENTS

	Page
1. Introduction	9
2. Some basic terms	11
2.1 Samples and Populations	11
2.2 Arithmetic mean	11
2.3 Variance	12
2.4 Statistics and parameters	13
3. Frequency distributions	14
3.1 Some terms associated with a frequency distribution	14
3.2 Four well-known frequency distributions	16
3.2.1 Positive binomial distribution ($\sigma^2 < \mu$)	17
3.2.2 Poisson series ($\sigma^2 = \mu$)	18
3.2.3 Negative binomial distribution ($\sigma^2 > \mu$)	23
3.2.4 Normal distribution and transformations	30
3.2.5 The binomial family	36
4. The spatial dispersion of a Population	37
4.1 Random distribution	38
4.1.1 A rapid test for agreement with a Poisson series	40
4.1.2 Variance to mean ratio	40
4.1.3 χ^2 test for "goodness of fit"	44
4.2 Regular distribution	46
4.2.1 The use of a positive binomial distribution as an approximate model for a regular distribution	46
4.3 Summary	48
5. Contagious distributions	50
5.1 The diversity of contagious distributions	50
5.2 Negative binomial distribution	51
5.2.1 Tests for agreement with a negative binomial. Large samples ($n > 50$)	53
5.2.2 Other tests for agreement with a negative binomial, and tests for small samples ($n < 50$)	54
5.2.3 Estimating a common k for a series of samples	63
5.3 Other contagious frequency distributions	66
5.4 Effect of quadrat size	68
5.5 Taylor's power law	71

	Page
5.6 Indices of dispersion	73
5.6.1 Indices based on the variance to mean ratio	74
5.6.2 k in the negative binomial	75
5.6.3 b of Taylor's power law	76
5.6.4 Morisita's index of dispersion	76
5.6.5 Other indices of dispersion	77
5.7 Summary	78
6. The precision of a sample mean	**80**
6.1 Standard error of the mean	80
6.2 Confidence limits of the mean	81
6.2.1 Normal approximation with large samples	81
6.2.2 Small samples ($n<30$) from a Poisson series (Random distribution)	83
6.2.3 Small samples ($n<30$) from a positive binomial distribution (Regular distribution)	86
6.2.4 Small samples ($n<30$) from a contagious distribution	86
6.3 Summary	92
7. Comparison of samples	**94**
7.1 Parametric tests	95
7.1.1 Methods associated with the normal distribution, and their application to large samples ($n>50$)	95
7.1.2 Samples from a Poisson Series (Random distribution)	103
7.1.3 Small samples ($n<50$) from contagious distributions	105
7.2 Non-parametric methods	112
7.3 Summary guide	124
8. Planning a sampling programme	**126**
8.1 Faunal surveys	126
8.2 Quantitative studies	127
8.2.1 The dimensions of the sampling unit (quadrat size)	128
8.2.2 The number of sampling units in each sample	128
8.2.3 The location of sampling units in the sampling area	131
8.3 Sub-sampling in the laboratory with large catches	135
8.4 Summary guide	136

		Page
9.	Acknowledgments	138
10.	References	139
	Appendix: symbols and terms	143
	1. Mathematical symbols	143
	2. Greek symbols	143
	3. Latin symbols	143

EXAMPLES

		Page
1.	Calculation of arithmetic mean	12
2.	Calculation of variance	13
3.	Arrangement of counts in a frequency distribution, and calculation of mean and variance from a frequency distribution	14
4.	Calculation of expected frequencies in a positive binomial distribution	18
5.	Calculation of expected frequences in a Poisson series	19
6.	Calculation of k and the expected frequencies in a negative binomial distribution	25
7.	A. Effect of a transformation on a negative binomial distribution	34
	B. Effect of a transformation on the relationship between mean and variance	34
8.	χ^2 test (variance to mean ratio) for agreement with a Poisson series; small samples ($n<31$)	41
9.	χ^2 test (variance to mean ratio) for agreement with a Poisson series; large samples ($n>31$)	43
10.	χ^2 test (goodness-of-fit) for agreement with a Poisson series	45
11.	Test for agreement with a positive binomial	47
12.	χ^2 test (goodness-of-fit) for agreement with a negative binomial distribution	53
13.	A. Method 2 for k, and statistic U for testing agreement with a negative binomial distribution	58
	B. χ^2 test for goodness-of-fit	60
14.	Method 1 for k, and statistic T for testing agreement with a negative binomial distribution	60
15.	Method 3 for k and statistic T for testing agreement with a negative binomial distribution	62
16.	Calculation of a common k	65

CONTENTS

	Page
17. Application of Taylor's power law	72
A. Graphical method	72
B. Calculation of a and b	73
18. Calculation of standard error and 95% confidence limits for a large sample ($n > 30$)	83
19. Calculation of 95% confidence limits for a small sample ($n < 30$) from a Poisson series	84
20. Calculation of 95% confidence limits for a small sample from a negative binomial	88
21. Calculation of 95% confidence limits for a small sample when the power law is applicable	89
22. A. Use of the logarithmic transformation to calculate 95% confidence limits	91
B. Use of the log $(x + 1)$ transformation to calculate 95% confidence limits	91
23. Comparison of means ($n > 50$)	96
24. Comparison of variances ($n > 50$)	97
25. Comparison of means of two samples from Poisson series ($nm > 30$)	104
26. A. Comparison of two small samples from contagious distributions (F-test and t-test)	106
B. Comparison of two small samples from contagious distribution (one-way analysis of variance)	107
27. A. Comparison of more than two samples from contagious distributions (one-way analysis of variance)	108
B. Checking the adequacy of the transformation	109
28. Comparison of more than two samples from contagious distributions (two-way analysis of variance)	111
29. Comparison of two small samples from contagious distributions (Mann-Whitney U-test)	115
30. Quenouille's test of the differences in mean level between several samples	116
31. Kruskal-Wallis test for differences in mean level between several samples	118

32. Friedman's test for differences in mean level between several samples 120
33. χ^2 test for a single classification ($k>2$) 122
34. Use of a 2 by 2 contingency table 123
35. Use of a large contingency table 123

I INTRODUCTION

The purpose of sampling the bottom fauna of freshwater habitats is usually to discover which species are present, and to detect spatial and temporal changes in the density of each species. First, the ecologist must choose a suitable method for removing quantitative samples of the benthos, and there are several guides to the numerous samplers (Welch 1948, Macan 1958, Albrecht 1959, Cummins 1962, and Southwood 1966). Next the catch must be identified, preferably to species, and there are now keys to many groups of aquatic invertebrates (see scientific publications of the F.B.A., and Macan 1959, for a "Key to other Keys"). Finally the catch must be counted and the ecologist must decide how to analyse the counts.

The purpose of the present guide is to give an elementary account of some suitable statistical methods. These methods provide information on the spatial distribution of the bottom fauna, they enable estimates to be made of the total population in an area of bottom, and they provide a sound basis for judging the significance of quantitative differences between samples. A knowledge of statistical methods is essential if experimental and survey techniques are to be correctly geared to the objectives of an investigation, and if the methods of testing hypotheses and making estimates are to be scientifically valid. Statistical methods should not be used as a salvage operation!

All methods are described in detail with a minimum of mathematical theory, and all data for the examples were collected from small streams in the Lake District. No attempt is made to offer a full bibliography and only "key-references" are given. The tables of Pearson & Hartley (1966) are essential and must be obtained before some methods can be applied. The laborious calculations of the more complex methods are considerably reduced by the use of tables, a calculating machine, or an electronic computer.

Chapter 2 introduces some basic statistical terms, and chapter 3 describes the theory and properties of some well-known mathematical distributions. A knowledge of these distributions is essential before (1) the spatial distribution of the bottom fauna can be described in mathematical terms (chapters 4 and 5); (2) errors of population parameters can be estimated (chapter 6); (3) temporal and spatial changes in density can be compared (chapter 7). Chapter 6 deals with the problems of estimating numbers per unit area and total numbers in a large section of bottom. Chapter 7 describes both parametric and non-parametric methods for the quantitative comparison of samples. Chapter 8 discusses the problems of planning a sampling programme. A brief summary guide is given at the end of each chapter, and a glossary of symbols and terms is included as an appendix.

II SOME BASIC TERMS

2.1 SAMPLES AND POPULATIONS

It is rarely possible to count all the invertebrates in a population and therefore samples have to be taken. The two definitions of the term *population* are distinguished in this account by a capital P for the ecological meaning and a small p for the statistical meaning. A "Population" in ecology is the total numbers of a species in a definite area, which may be the whole bottom of a lake or river, or only a section of lake shore or river. In statistical terminology, any aggregate of values is termed a "population", and therefore the whole aggregate of sampling units into which an area is divided is known as the "population of sampling units". For example, the population for a lake or river is the total area of bottom and this area is divided into sampling units of equal size. Each action of a sampler removes one sampling unit from the bottom, and the size of the sampling unit depends upon the type of sampler used. The sampling units must be distinct and non-overlapping, and they must together constitute the whole of the population. As it is rarely possible to remove all the sampling units in the population, a group of units is selected from the whole aggregate and is thought to be representative of the whole population. This group of sampling units forms a *sample*. The units in the sample are of equal size, are taken within a short period of time, and are usually selected at random from the whole population of sampling units (see section 8.2.3). The basic information is always obtained by counting the individuals in each sampling unit of the sample. Therefore the number of individuals per sampling unit is the variable under study.

2.2 ARITHMETIC MEAN

If the counts for a series of sampling units are represented by $x_1, x_2, x_3, \ldots x_n$, and there are n sampling units in a sample, then the sum of the counts is $x_1 + x_2 + x_3 \ldots + x_n$ or Σx in a shorter form, where Σ means "sum of" and thus Σx is "sum of x". The "arithmetic mean" of the sample is represented by $\bar{x}$ (called "bar x") and is given by:

$$\bar{x} = \frac{\Sigma x}{n}$$

There are several different "averages" for a series of counts, *e.g.* median, mode, arithmetic mean, harmonic mean, and geometric mean, and it should always be clearly stated which average value has been calculated. The arithmetic mean is most frequently used in this account, but the geometric mean (see section 6.2.4) and median (see section 7.2) are also used.

Example 1. Calculation of arithmetic mean

A Macan shovel sampler (described in Macan 1958) was used to take a random sample of 11 sampling units from the bottom of a stony stream. The sampler removes a square of bottom equal to 500 cm^2 (0·05 m^2). Nymphs of the mayfly *Baëtis rhodani* were counted in each sampling unit and the following values were obtained:

$$14, 15, 12, 7, 8, 14, 11, 14, 10, 9, 10$$

Therefore the sum of these counts (Σx) is:

$$14 + 15 + 12 + \cdots + 10 = 124,$$

and the number of sampling units (n) in the sample is 11. The arithmetic mean of the sample ($\bar{x}$) is given by:

$$\bar{x} = \frac{\Sigma x}{n} = \frac{124}{11} = 11\cdot273$$

Therefore the mean number of nymphs in 500 cm^2 is 11·27.

2.3 Variance

There is always some variation between the counts of a sample, and this variation will be partly due to inadequacies in the sampling technique, *e.g.* some invertebrates may pass through the collecting net, escape from the sampler, or be missed by the sampler (discussed by Macan 1958, Albrecht 1959, Cummins 1962). It is often impossible to assess the *error* of a sampler, but this error will usually be of the same magnitude for all the units in a sample. If the sampler is reasonably efficient, the variation between counts will be chiefly due to the spatial distribution of the invertebrates on the bottom of the lake or stream. This variation can be expressed in several ways. The *range* is the simplest and is the difference between the highest and lowest counts, *e.g.* range = 15 − 7 = 8 for the counts of example 1. As the range depends solely on the extreme counts, it is of limited value. Each count differs from the arithmetic mean of the sample by a quantity called the *deviation* and the *variance* of the sample (s^2) is simply the mean of the squares of the deviations thus:

$$s^2 = \frac{\Sigma(x-\bar{x})^2}{n}$$

2.4 STATISTICS AND PARAMETERS

As the sample variance is often an estimate of the variance of the population, the equation is usually written thus:

$$s^2 = \frac{\Sigma(x-\bar{x})^2}{n-1} = \frac{\Sigma(x^2)-(\Sigma x)^2/n}{n-1} = \frac{\Sigma(x^2)-\bar{x}\Sigma x}{n-1}$$

where $n-1$ is the *degrees of freedom* of the sample. These three equations give the same answer, but the calculations involved in the first version are often very laborious. The *standard deviation* of the sample (s) is the square-root of the variance ($s = \sqrt{s^2}$). A complete explanation of degrees of freedom is long and involved, but the -1 can be regarded as a tax for using the sample mean $\bar{x}$ (a statistic), instead of the population mean μ (a parameter), in the estimate of the population variance σ^2 (sample variances tend to underestimate population variance).

Example 2. Calculation of variance

Counts as in example 1 and therefore $\Sigma x = 124$, $\bar{x} = 11\cdot273$, and $n-1 = 11-1 = 10$. First calculate

$$\Sigma(x^2) = 14^2 + 15^2 + 12^2 + \cdots + 10^2 = 1472$$

and then s^2 from the equation

$$s^2 = \frac{\Sigma(x^2)-\bar{x}\Sigma x}{n-1} = \frac{1472-11\cdot273\,(124)}{10} = 7\cdot415$$

2.4 STATISTICS AND PARAMETERS

We can now describe a sample in terms of an average count and also a measure of the spread of the counts, *i.e.* the arithmetic mean and variance of the sample. As both these values are calculated from the sample, they are called *statistics*. Each statistic may be used as an estimate of a population value and the latter is called a *parameter*. Parameters are usually represented by Greek letters and statistics by Roman letters (Table 1).

TABLE 1. COMPARISON OF SYMBOLS FOR PARAMETERS AND STATISTICS

	Statistics of sample	Parameters of population
Arithmetic mean	$\bar{x}$	μ (pronounced mu)
Variance	s^2	σ^2 (sigma-squared)
Standard deviation	s	σ (sigma)
Mean and variance of Poisson	m	λ (lambda)
Number of sampling units	n	

III FREQUENCY DISTRIBUTIONS

3.1 Some terms associated with a frequency distribution

A variable quantity can either be *continuous*, *i.e.* it can assume any value within a certain range; or be *discontinuous*, *i.e.* it can only assume integral values (whole numbers) and not fractions of integers. Continuous variables are usually measurements, *e.g.* heights, lengths and weights of animals; whereas discontinuous variables are usually counts, *e.g.* number of petals on a flower, number of hairs on an insect, number of invertebrates in a sampling unit.

If a large number of measurements or counts of a variable are made, the figures can be summarised in a frequency distribution. The figures are first placed in numerical order and then grouped into *frequency classes*. As counts from bottom samples are always discontinuous, each class is usually an integer and the number of values that fall into a class is the *class frequency*. Therefore the frequency simply records the number of sampling units which contain the same number of invertebrates. If each class covers more than an integer, the classes should be non-overlapping and the class-intervals should be equal, *e.g.* 1 to 5, 6 to 10, 11 to 15, etc. with class-interval of 5.

The form or pattern of a frequency distribution is shown by the distribution in numerical form but is more clearly recognised in a diagram such as a *histogram* (see Fig. 1). A histogram is a block diagram in which each frequency class is represented by a column. The area of each column is proportional to the frequency and if the bases of the columns are equal (*i.e.* the class-intervals are the same), then the height of each column is also proportional to frequency. Further examples of histograms are given in Figs. 2–4.

Example 3. Arrangement of counts in a frequency distribution, and calculation of mean and variance from a frequency distribution

A large random sample was taken of the number of mayfly nymphs (*Baëtis rhodani*) on the tops of stones at night. There were 80 sampling units in the sample and each sampling unit was located by the use of random co-ordinates (see section 8). The following counts were obtained of the number of nymphs in each sampling unit of 100 cm^2 (10 cm by 10 cm quadrat): 9, 8, 11, 11, 14, 6, 17, 8, 13, 11, 6, 9, 13, 8, 10, 13, 10, 7, 12, 8, 16, 4, 11, 11, 5, 15, 9, 14, 12, 15, 9, 7, 10, 10, 12, 7, 12, 8, 9, 10, 7, 11, 16, 7, 13, 6, 10, 11, 8, 9, 11, 8, 12, 13, 4, 15, 9, 11, 10, 9, 14, 5, 8, 12, 12, 10, 7, 8, 7, 11, 9, 6, 10, 12, 13, 8, 14, 9, 10, 15. Many of the

3.1 SOME TERMS ASSOCIATED WITH A FREQUENCY DISTRIBUTION

counts are the same for several of the sampling units, e.g. a count of 8 occurs in 10 sampling units and therefore the frequency of the class 8 is 10. The whole frequency distribution is:

$$f \quad 2 \quad 2 \quad 4 \quad 7 \quad 10 \quad 10 \quad 10 \quad 10 \quad 8 \quad 6 \quad 4 \quad 4 \quad 2 \quad 1$$
$$\Sigma f = n = 80$$
$$x \quad 4 \quad 5 \quad 6 \quad 7 \quad 8 \quad 9 \quad 10 \quad 11 \quad 12 \quad 13 \quad 14 \quad 15 \quad 16 \quad 17$$

where x is a particular count (number of invertebrates in a sampling unit) and f is the frequency of that particular count in the sample. The distribution is also shown as a histogram in Fig. 1. As f is also the number of sampling units with the same count, the total number of sampling units (n) in the sample is $\Sigma f = 80$.

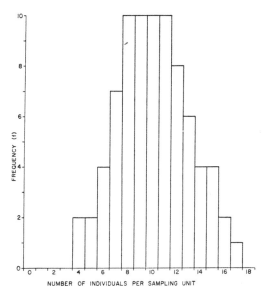

FIGURE 1. Frequency distribution of counts from Example 3.

Note that x now represents a count which is common to several sampling units, whereas x represented the count for *each* sampling unit in examples 1 and 2. The total number of invertebrates in the sample is Σfx and therefore the arithmetic mean ($\bar{x}$) of the sample is given by:

$$\bar{x} = \frac{\Sigma fx}{n} = \frac{810}{80} = 10 \cdot 125$$

This is an alternative method to that given in example 1 and both methods give the same value for the mean (a check is to use the method of example 1 with the 80 separate counts).

The variance (s^2) of the sample is given by:

$$s^2 = \frac{\Sigma(fx^2) - \bar{x}\Sigma fx}{n-1}$$

This is an alternative method to that given in example 2.

$$\Sigma(fx^2) = 2(4^2) + 2(5^2) + 4(6^2) + \cdots + 1(17^2) = 8880$$

Therefore

$$s^2 = \frac{8880 - 10 \cdot 125\,(810)}{79} = 8 \cdot 5918$$

3.2 Four well-known frequency distributions

Known mathematical frequency distributions can be used as models for samples from a population. If the frequency distribution of counts from a sample fits one of these models, then:

(1) the *spatial dispersion of the Population* (ecological meaning) can be described in mathematical terms (sections 4 and 5),
(2) errors of population parameters can be estimated (section 6),
(3) temporal and spatial changes in density can be compared (section 7),
(4) the effect of environmental factors can be assessed.

The following mathematical distributions are often suitable models for the three possible relationships between the variance (σ^2) and arithmetic mean (μ) of a population:

(1) *Positive binomial:* this is an approximate model when the variance is significantly less than the mean ($\sigma^2 < \mu$).
(2) *Poisson series:* this is the only model when the variance is approximately equal to the mean ($\sigma^2 = \mu$).
(3) *Negative binomial:* this is the most flexible of several possible models when the variance is significantly greater than the mean ($\sigma^2 > \mu$).

These three models are strictly probability distributions and are converted to frequency distributions by multiplying each probability by the sample size (see examples 4, 5, and 6). The three models, plus the normal distribution, will now be described.

3.2 FOUR WELL-KNOWN FREQUENCY DISTRIBUTIONS

3.2.1 Positive binomial distribution ($\sigma^2 < \mu$).

The positive binomial is the basis of the Binomial family of distributions and is best introduced through the concept of *probability*.

If there are 9 white balls and 1 black ball in a bag, then the chance of a ball picked out of the bag at random being black is clearly 1 in 10. This is written as $p = 0.1$ and similarly the probability (or chance) of the ball being white is 9 in 10 or $p = 0.9$. If all the balls were white, then the probability of a ball picked out at random being white is 1 ($p = 1$) and being black is 0 ($p = 0$). Therefore the use of the concept of probability answers the question "What is the chance that a stated event will occur?", and the probability scale ranges from $p = 0$ for impossible events to $p = 1$ for certain events.

In the positive binomial, p is the probability of an event occurring one way and q is the probability of the same event occurring in an alternative way. If q and p remain constant in each of k independent trials (k is the number of times the situation is repeated) and $q+p = 1$, then the probability series is given by the expansion of $(q+p)^k$. The individual terms of the binomial expansion are given by:

$$P_{(x)} = \frac{k!}{x!(k-x)!} q^{k-x} p^x$$

where $P_{(x)}$ is the probability of x individuals in a sampling unit, and $k!$ means k factorial (*e.g.* 4! means $(4) \times (3) \times (2) \times (1)$). The population parameters are estimated thus; arithmetic mean μ by kp, and variance σ^2 by kpq, where k, p and q are sample statistics.

The positive binomial is usually used as a model when each individual in a sample can be recognised as having, or not having, a certain character, *e.g.* the proportion of sampled individuals which are male. A full description of this application of the positive binomial is given in most statistical textbooks (*e.g.* Bailey 1959, Snedecor & Cochran 1967).

When the positive binomial is used as an approximate model for bottom samples (with $s^2 < \bar{x}$), the following definitions (from Greig-Smith 1964) are applicable: k is the maximum possible number of individuals a sampling unit could contain, p is the probability of any one of the possible places in the sampling unit being occupied by an individual, and q is the probability of any one of the possible places not being occupied by an individual ($q = 1-p$). The expected probabilities are given by the expansion of $(q+p)^k$, and can be calculated (example 4) or obtained from tables. Table 37 in Pearson & Hartley (1966) gives individual terms of selected binomial series, and more complete tables are published by National Bureau of Standards (1950). Expected frequencies are calculated by multiplying each expected probability by the total number of sampling units.

As few bottom samples have a variance significantly less than the mean, the positive binomial is seldom required as a model. The special applications of the model, the estimation of k, and the comparison of observed and expected frequencies are discussed in section 4.2.

Example 4. Calculation of expected frequencies in a positive binomial distribution

In a sample of 20 sampling units ($n = 20$), the arithmetic mean is 3 ($\bar{x} = 3$), and each sampling unit could contain a maximum of 4 individuals ($k = 4$).

$$p = \frac{\bar{x}}{k} = \frac{3}{4} = 0.75, \qquad q = 1 - p = 0.25$$

Expected probabilities are given by the expansion of:

$$(q+p)^k = (0.25 + 0.75)^4$$

The probability of 0 individuals per sampling unit is

$$P_{(x=0)} = \frac{4!}{0!(4-0)!} q^{4-0} p^0 = q^4 = 0.25^4 = 0.0039$$

The probability of 1 individual per sampling unit is

$$P_{(x=1)} = \frac{4!}{1!(4-1)!} q^{4-1} p^1 = 4q^3 p = 4(0.25)^3 0.75 = 0.0469$$

The probabilities of 2, 3, 4 individuals per sampling unit are

$$6q^2 p^2 = 0.2109, \qquad 4qp^3 = 0.4219, \qquad p^4 = 0.3164.$$

Note that total probability is 1, *i.e.*

$$q^4 + 4q^3 p + 6q^2 p^2 + 4qp^3 + p^4$$
$$= 0.0039 + 0.0469 + 0.2109 + 0.4219 + 0.3164 = 1.0000.$$

Expected frequencies for counts of 0, 1, 2, 3, 4, are

20(0·0039), 20(0·0469), 20(0·2109), 20(0·4219), 20(0·3164),

The whole frequency distribution is:

f	0·08	0·94	4·22	8·44	6·33
x	0	1	2	3	4

These expected frequencies are compared with the observed frequencies in example 11.

3.2.2 *Poisson series* ($\sigma^2 = \mu$)

In the positive binomial distribution, the arithmetic mean $\mu = kp$, and the variance $\sigma^2 = kpq = \mu q$. As $q = 1 - p = 1 - \mu/k$, then

$$\sigma^2 = \mu q = \mu \left(1 - \frac{\mu}{k}\right) = \mu - \frac{\mu^2}{k}.$$

3.2 FOUR WELL-KNOWN FREQUENCY DISTRIBUTIONS

Therefore, if k is finite, the variance will be always less than the mean ($\sigma^2 < \mu$). As k increases and approaches infinity, μ^2/k approaches 0 and σ^2 approaches μ in the equation $\sigma^2 = \mu - \mu^2/k$. Eventually $\sigma^2 = \mu$ and then the probability series follows the discontinuous Poisson series:

$$P_{(x)} = e^{-\lambda}\frac{\lambda^x}{x!}$$

where $P_{(x)}$ is the probability of x individuals in a sampling unit, λ is the Poisson parameter ($\lambda = \mu = \sigma^2$), e is the base of natural (Napierian) logarithms (e = 2·7183), and $x!$ means x factorial.

Whereas two parameters (k, p) are needed to determine the probability series of the positive binomial, only one parameter (λ) is required for the Poisson probability series. Therefore it is sufficient to know only kp ($kp = \mu = \lambda$) and not k and p separately. The Poisson parameter λ is estimated by the statistic m ($m = \bar{x} = s^2$) and thus:

$$P_{(x)} = e^{-m}\frac{m^x}{x!}$$

i.e. The probabilities of $0, 1, 2, 3, \ldots x$ individuals per sampling unit are

$$e^{-m}, \ e^{-m}\frac{m^1}{1!}, \ e^{-m}\frac{m^2}{2!}, \ e^{-m}\frac{m^3}{3!}, \ \ldots \ e^{-m}\frac{m^x}{x!}.$$

Table 39 in Pearson & Hartley (1966) gives individual terms of the Poisson series for different values of m from 0·1 to 15·0. Alternatively, the individual terms can be calculated (see Example 5). The expected frequency of each count is given by $nP_{(x)}$, *i.e.* by multiplying each term of the series by the total number of sampling units (n).

Example 5. Calculation of expected frequencies in a Poisson series

The arithmetic mean of a sample of 80 sampling units is 10·125. Therefore $\bar{x} = 10 \cdot 125 = m$ and $n = 80$. The probability of 0 individuals per sampling unit is:

$$P_{(x=0)} = e^{-10 \cdot 125} = 2 \cdot 7183^{-10 \cdot 125}$$
$$\log P_{(x=0)} = -10 \cdot 125 \log e = -(10 \cdot 125)(0 \cdot 43429)$$
$$= -4 \cdot 39719 \quad = -5 + 0 \cdot 60281$$

thus $\quad P_{(x=0)} = \text{antilog } \bar{5} \cdot 60281 = 0 \cdot 000040068$

The expected frequency of 0 individuals per sampling unit is

$$n \, e^{-10 \cdot 125} = 80(0 \cdot 000040068) = 0 \cdot 0032054 \simeq 0$$

The remaining probabilities are easily calculated by the successive terms of the Poisson series:

$$P_{(x=1)} = e^{-m}\frac{m^1}{1!} = (0 \cdot 000040068)(10 \cdot 125) = 0 \cdot 00040569$$

TABLE 2. INDIVIDUAL TERMS OF THE POISSON SERIES

for $m=10\cdot125$, and frequency distribution for a sample of 80 sampling units ($n=80$). x is a particular count, $P_{(x)}$ is the probability of x individuals in a sampling unit, and f is the frequency of a particular count in the sample (f=number of sampling units with the same count).

Probabilities and expected frequencies are given to 5 significant figures.

x	$P_{(x)}$	$f=nP_{(x)}$	f (as integer)
0	$e^{-m}=0\cdot000040068$	0·0032054	0
1	$e^{-m}m=0\cdot00040569$	0·032455	0
2	$e^{-m}\dfrac{m^2}{2!}=0\cdot0020538$	0·16430	0
3	$e^{-m}\dfrac{m^3}{3!}=0\cdot0069316$	0·55453	1
4	$e^{-m}\dfrac{m^4}{4!}=0\cdot017546$	1·4037	1
5	$e^{-m}\dfrac{m^5}{5!}=0\cdot035530$	2·8424	3
6	$e^{-m}\dfrac{m^6}{6!}=0\cdot059957$	4·7966	5
7	$e^{-m}\dfrac{m^7}{7!}=0\cdot086723$	6·9379	7
8	$e^{-m}\dfrac{m^8}{8!}=0\cdot10975$	8·7807	9
9	$e^{-m}\dfrac{m^9}{9!}=0\cdot12347$	9·8775	10
10	$e^{-m}\dfrac{m^{10}}{10!}=0\cdot12501$	10·0008	10
11	$e^{-m}\dfrac{m^{11}}{11!}=0\cdot11507$	9·2053	9
12	$e^{-m}\dfrac{m^{12}}{12!}=0\cdot097090$	7·7672	8

3.2 FOUR WELL-KNOWN FREQUENCY DISTRIBUTIONS

TABLE 2—*continued*

13	$e^{-m}\dfrac{m^{13}}{13!}=0\cdot075618$	6·0496	6
14	$e^{-m}\dfrac{m^{14}}{14!}=0\cdot054688$	4·3752	4
15	$e^{-m}\dfrac{m^{15}}{15!}=0\cdot036914$	2·9536	3
16	$e^{-m}\dfrac{m^{16}}{16!}=0\cdot023360$	1·8691	2
17	$e^{-m}\dfrac{m^{17}}{17!}=0\cdot013913$	1·1130	1
18	$e^{-m}\dfrac{m^{18}}{18!}=0\cdot0078259$	0·62607	1
19	$e^{-m}\dfrac{m^{19}}{19!}=0\cdot0041703$	0·33362	0
20	$e^{-m}\dfrac{m^{20}}{20!}=0\cdot0021112$	0·16890	0
21	$e^{-m}\dfrac{m^{21}}{21!}=0\cdot0010179$	0·081432	0
22	$e^{-m}\dfrac{m^{22}}{22!}=0\cdot00046846$	0·037477	0
23	$e^{-m}\dfrac{m^{23}}{23!}=0\cdot00020622$	0·016498	0
24	$e^{-m}\dfrac{m^{24}}{24!}=0\cdot000086999$	0·0069599	0
25	$e^{-m}\dfrac{m^{25}}{25!}=0\cdot000035235$	0·0028188	0
26	$e^{-m}\dfrac{m^{26}}{26!}=0\cdot000013721$	0·0010977	0

$$\Sigma P_{(x)}=1\cdot0000 \quad \Sigma nP_{(x)}=80\cdot002 \quad \Sigma f=80$$

$$P_{(x=2)} = e^{-m}\frac{m^2}{2!} = e^{-m}\frac{m}{1}\cdot\frac{m}{2} = (0\cdot00040569)\frac{(10\cdot125)}{2} = 0\cdot0020538$$

$$P_{(x=3)} = e^{-m}\frac{m^3}{3!} = e^{-m}\frac{m^2}{2!}\cdot\frac{m}{3} = (0\cdot0020538)\frac{(10\cdot125)}{3} = 0\cdot0069316$$

This process is continued until the individual terms are close to 0 (see Table 2). The expected frequencies of the counts are obtained by multiplying each probability by 80. These expected frequencies are compared with the observed frequencies in example 10.

The Poisson series is associated with events which occur *randomly* in a continuum of time or space, *e.g.* distribution of number of cars passing a given point per minute (for a large number of minutes) at a given time of day; distribution of number of telephone calls received at a given switchboard per minute (for a large number of minutes) for a given part of the day; distribution of counts of aquatic invertebrates per sampling unit (for a large number of units) in a bottom sample. Therefore tests for agreement with a Poisson series are used as tests for randomness of distribution, but it must be remembered that the Poisson series is only a special case of the positive binomial where p tends to 0 as k approaches infinity (because $p = \mu/k$) and q tends to 1 as σ^2 approaches μ (because $q = kpq/kp = \sigma^2/\mu = 1$).

The use of the Poisson series as a mathematical model for the distribution of counts (number of aquatic invertebrates per sampling unit) from a bottom sample involves the following conditions:

(1) The probability of any given point in the sampling area being occupied by a particular individual is constant and very small (constant $p \to 0$), and consequently there is a very high probability of any given point not being occupied by an individual (constant $q \to 1$).

(2) The number of individuals per sampling unit must be well below the maximum possible number that could occur in the sampling unit ($k \to \infty$).

(3) The presence of an individual at a given point must not increase or decrease the probability of another individual occurring near by: *i.e.* the individuals must be separate discrete units.

(4) The samples must be small relative to the population.

The first condition implies that there is an equal chance of an individual occupying any point in the sampling area and this condition is fulfilled if the individuals are distributed completely at random on the bottom. If the individuals of a species are relatively crowded in the sampling area, the number of individuals per sampling unit approaches the maximum possible and the second condition is not fulfilled. Under these conditions the variance is less than the mean of the sample ($s^2 < \bar{x}$) and the positive binomial is a more suitable model for the distribution of counts. Non-fulfilment of the third condition indicates a non-random distribution of the aquatic invertebrates on the bottom, and therefore the Poisson series is no longer applicable. The removal of a sampling unit from a finite population must affect the value of p for the next sampling unit, but this effect will be minute if a very small proportion of the population is removed in each sampling unit. Therefore the fourth condition ensures that the value of p does not alter significantly from one trial to the next.

3.2.3 *Negative binomial distribution* ($\sigma^2 > \mu$)

If the first and third conditions for the use of the Poisson series are not fulfilled, the variance of the population is usually greater than the arithmetic mean ($\sigma^2 > \mu$) and the population is clumped or aggregated. Several mathematical distributions have been used as models for this situation and the negative binomial distribution is often a suitable model for invertebrate populations (Anscombe 1949, Bliss & Fisher 1953, Debauche 1962).

The negative binomial is the mathematical counterpart of the positive binomial and therefore the probability series of the negative binomial is given by the expansion of $(q-p)^{-k}$, where $p = \mu/k$ and $q = 1+p$. The parameters of this distribution are the arithmetic mean μ and the exponent k. Note that k is no longer the maximum possible number of individuals a sampling unit could contain, but is related to the spatial distribution of the aquatic invertebrates on the bottom. Unlike the positive binomial, k is not necessarily an integer in the negative binomial. The variance of the population $\sigma^2 = kpq = \mu q = \mu(1+\mu/k) = \mu + \mu^2/k$. Therefore the reciprocal of the exponent k, i.e. $1/k$, is a measure of the excess variance or *clumping* of the individuals in a population. As $1/k$ approaches 0 and k approaches infinity, the distribution converges to the Poisson series ($\sigma^2 \to \mu$). Conversely, if clumping increases, $1/k$ approaches infinity ($k \to 0$) and the distribution converges to the Logarithmic series (Fisher, Corbet & Williams 1943). A full description and many examples of the application of the logarithmic series are given in the book by Williams (1964). Therefore the limiting values of k lead to other important mathematical distributions.

The individual terms of $(q-p)^{-k}$ are given by:

$$P_{(x)} = \left(1+\frac{\mu}{k}\right)^{-k} \frac{(k+x-1)!}{x!(k-1)!} \left(\frac{\mu}{\mu+k}\right)^x$$

where $P_{(x)}$ is the probability of x individuals in a sampling unit. Hence the expected frequency of a particular count is $nP_{(x)}$ (as in the positive binomial and Poisson series) where n is the total number of sampling units in the sample.

The parameters μ and k are estimated from the frequency distribution of the sample by the statistics $\bar{x}$ and $\hat{k}$. The arithmetic mean $\bar{x}$ is calculated in the usual way (see examples 1 and 3). There are several methods of calculating $\hat{k}$ (Anscombe 1949, 1950, Bliss & Fisher 1953, Debauche 1962). Most of these methods are approximate and should only be used as the first step towards the more accurate method of *maximum likelihood*. A very simple, but approximate, method is derived from the equation for the variance of the negative binomial:

$$\sigma^2 = \mu + \frac{\mu^2}{k} \text{ and therefore } k = \frac{\mu^2}{\sigma^2 - \mu}$$

The parameters μ and σ^2 are estimated by the statistics $\bar{x}$ and s^2 in the usual way and therefore:

$$\hat{k} = \frac{\bar{x}^2}{s^2 - \bar{x}}$$

This method is not very efficient for values of k below 4, unless $\bar{x}$ is also less than 4 (Anscombe, 1950, gives a diagram showing the efficiency of the estimation of k by this method). The method does provide a rough estimate of $\hat{k}$ for substitution in the maximum-likelihood equation:

$$n\log_e\left(1+\frac{\bar{x}}{\hat{k}}\right) = \sum\left(\frac{A_{(x)}}{\hat{k}+x}\right)$$

where n is the total number of sampling units, $\log_e$ designates a natural (Napierian) logarithm [$\log_e x = (\log_{10} x)(\log_e 10) = (\log_{10} x)(2\cdot30259)$, *i.e.* (common logarithm x) ($2\cdot30259$)], and $A_{(x)}$ is the total number of counts exceeding x. Different values of $\hat{k}$ are tried until the equation is balanced, *i.e.* the equation is solved by iteration (trial and error).

3.2 FOUR WELL-KNOWN FREQUENCY DISTRIBUTIONS

Example 6. Calculation of $\hat{k}$ and the expected frequencies in a negative binomial distribution

The frequency distribution of counts from a random sample of 80 sampling units is:

x	0	1	2	3	4	5	6	7	8	9	10	11	12	13	14	15	16
f	3	7	9	12	10	6	7	6	5	4	3	2	2	1	1	1	1
$A_{(x)}$	77	70	61	49	39	33	26	20	15	11	8	6	4	3	2	1	0

where x is a particular count, f is the frequency of that particular count, and $A_{(x)}$ is the total number of counts exceeding x. Total number of sampling units $n = \Sigma f = 80$, $\Sigma fx = 425$,

Arithmetic mean of sample $\bar{x} = \dfrac{\Sigma fx}{n} = \dfrac{425}{80} = 5\cdot 3125$

Variance of sample $s^2 = \dfrac{\Sigma(fx^2) - \bar{x}\Sigma fx}{n-1} = \dfrac{3327 - 5\cdot 3125(425)}{80 - 1} = 13\cdot 534$

A rough estimate of $\hat{k}$ is:

$$\hat{k} = \dfrac{\bar{x}^2}{s^2 - \bar{x}} = \dfrac{5\cdot 3^2}{13\cdot 5 - 5\cdot 3} = 3\cdot 4$$

As $\hat{k}$ probably lies close to 3·4, we try $\hat{k} = 3\cdot 3$ and $\hat{k} = 3\cdot 5$ in the maximum-likelihood equation, and then find an accurate estimate of k by proportion. Therefore $\hat{k} = 3\cdot 3$ is a suitable value to substitute in the likelihood equation:

$$n \log_e \left(1 + \dfrac{\bar{x}}{\hat{k}}\right) = \sum \left(\dfrac{A_{(x)}}{\hat{k} + x}\right)$$

Each side of the equation is solved thus:

$$n \log_e \left(1 + \dfrac{\bar{x}}{\hat{k}}\right) = n \log_{10} \left(1 + \dfrac{\bar{x}}{\hat{k}}\right) \log_e 10$$

$$= 80 \log_{10} \left(1 + \dfrac{5\cdot 3125}{3\cdot 3}\right) 2\cdot 30259$$

$$= 80(0\cdot 41664) \, 2\cdot 30259$$

$$= 76\cdot 7480$$

$$\sum \left(\dfrac{A_{(x)}}{\hat{k} + x}\right) = \dfrac{A_{(x=0)}}{\hat{k}} + \dfrac{A_{(x=1)}}{\hat{k}+1} + \dfrac{A_{(x=2)}}{\hat{k}+2} + \dfrac{A_{(x=3)}}{\hat{k}+3} \cdots \dfrac{A_{(x=15)}}{\hat{k}+15}$$

$$= \dfrac{77}{3\cdot 3} + \dfrac{70}{4\cdot 3} + \dfrac{61}{5\cdot 3} + \dfrac{49}{6\cdot 3} \cdots \dfrac{1}{18\cdot 3}$$

$$= 76\cdot 8200$$

Therefore the difference between the two sides of the likelihood equation is $76\cdot 7480 - 76\cdot 8200 = -0\cdot 0720$.

The difference will be 0 when the two sides balance and next we try $\hat{k} = 3\cdot 5$

$$80 \log_{10} \left(1 + \dfrac{5\cdot 3125}{3\cdot 5}\right) 2\cdot 30259 = 73\cdot 8728$$

FREQUENCY DISTRIBUTIONS

$$\sum \left(\frac{A_{(x)}}{3\cdot 5+x}\right) = 73\cdot 7000$$

Difference $= 73\cdot 8728 - 73\cdot 7000 = +0\cdot 1728$. Therefore the true value of $\hat{k}$ lies between 3·3 and 3·5, and is found by proportion:

k	Difference
3·3	$-0\cdot 0720$
$\hat{k}$	0
3·5	$+0\cdot 1728$

Therefore
$$\frac{\hat{k}-3\cdot 3}{3\cdot 5-3\cdot 3} = \frac{0\cdot 0720}{0\cdot 1728+0\cdot 0720}$$

$$\hat{k} = \frac{0\cdot 0720(0\cdot 2)+3\cdot 3}{0\cdot 1728+0\cdot 0720}$$

$$= 3\cdot 3588$$

or
$$\frac{3\cdot 5-\hat{k}}{3\cdot 5-3\cdot 3} = \frac{0\cdot 1728}{0\cdot 1728+0\cdot 0720}$$

$$\hat{k} = 3\cdot 5 - \frac{0\cdot 1728(0\cdot 2)}{0\cdot 1728+0\cdot 0720}$$

$$= 3\cdot 3588$$

Therefore $\bar{x} = 5\cdot 3125$ and $\hat{k} = 3\cdot 3588$ in the calculation of the expected probabilities and frequencies.

The probability of 0 individuals per sampling unit is:

$$P_{(x=0)} = \left(1+\frac{\bar{x}}{k}\right)^{-k} \frac{(k+0-1)!}{0!(k-1)!} \left(\frac{\bar{x}}{\bar{x}+k}\right)^0 = \left(1+\frac{\bar{x}}{k}\right)^{-k}$$

$$\log P_{(x=0)} = -k \log\left(1+\frac{\bar{x}}{k}\right) = -3\cdot 3588 \log\left(1+\frac{5\cdot 3125}{3\cdot 3588}\right)$$

$$= -1\cdot 38369 = \bar{2}+0\cdot 61631$$
$$P_{(x=0)} = \text{antilog } \bar{2}\cdot 61631 = 0\cdot 04133$$

Therefore the expected frequency for a count of $0 = nP_{(x=0)} = 80(0\cdot 04133) = 3\cdot 31$. Probability of 1 individual per sampling unit is:

$$P_{(x=1)} = \left(1+\frac{\bar{x}}{k}\right)^{-k} \frac{(k+1-1)!}{1!(k-1)!} \left(\frac{\bar{x}}{\bar{x}+k}\right)^1$$

$$= (P_{(x=0)}) \frac{k}{1} \left(\frac{\bar{x}}{\bar{x}+k}\right)^1$$

$$= (0\cdot 0413)(3\cdot 3588)\left(\frac{5\cdot 3125}{8\cdot 6713}\right)$$

$$= 0\cdot 0851$$
$$nP_{(x=1)} = 80(0\cdot 0851) = 6\cdot 81$$

3.2 FOUR WELL-KNOWN FREQUENCY DISTRIBUTIONS

Probability of 2 individuals per sampling unit is:

$$P_{(x=2)} = (P_{(x=1)}) \frac{(k+2-1)}{2} \left(\frac{\bar{x}}{\bar{x}+k}\right)$$

$$= (0 \cdot 0851) \frac{(3 \cdot 3588+1)}{2} \left(\frac{5 \cdot 3125}{8 \cdot 6713}\right)$$

$$= 0 \cdot 1136$$

$$nP_{(x=2)} = 80(0 \cdot 1136) = 9 \cdot 09$$

This process is continued until $\Sigma P_{(x)} \simeq 1$ and $\Sigma f \simeq 80$ (see Table 3). In theory, the total probability should always be 1, but there is always a small discrepancy due to rounding off the last place of decimals. There is close agreement between the expected frequencies (Table 3) and the actual frequencies given at the beginning of the example. Therefore the negative binomial distribution appears to be a good fit to the original counts, and this "goodness of fit" can be tested by χ^2 or by comparing actual and expected moments (see section 5.2).

Williamson & Bretherton (1963) give tables of expected probabilities for 1480 negative binomial distributions. These tables cover values of the parameter k from 0·1 to 200 (k always given to nearest 0·1), and are arranged in increasing size of a second parameter "p". The latter parameter is equivalent to $1/q$ and *not* p in the present account. This discrepancy arises because Williamson & Bretherton write the negative binomial distribution in the form $p^k(1-q)^{-k}$ where $q+p=1$. Therefore $\mu = kq/p$; $\sigma^2 = kq/p$; and the probability of 0 individuals per sampling unit is now p^k (instead of $q^{-k} = (1+\mu/k)^{-k}$ in the present account). An estimate of "p" is given by:

$$\frac{1}{1+\bar{x}/k}$$

e.g. for counts of example 6, $\bar{x} = 5 \cdot 31$, $s^2 = 13 \cdot 53$, and $k = 3 \cdot 4$ (to nearest 0·1). Therefore

$$\text{``}p\text{''} = \frac{1}{1+\bar{x}/k} = \frac{1}{1+5 \cdot 31/3 \cdot 4} = 0 \cdot 39$$

The nearest table in Williamson & Bretherton is that for "p" = 0·40, $k = 3 \cdot 5$, and the individual probabilities given in the table are very similar to those calculated in example 6.

TABLE 3. INDIVIDUAL TERMS OF THE NEGATIVE BINOMIAL

for $\bar{x} = 5\cdot3125$ and $\hat{k} = 3\cdot3588$; and frequency distribution for a sample of 80 sampling units ($n = 80$). x is a particular count, $P_{(x)}$ is the probability of x individuals in a sampling unit, and f is the frequency of a particular count in the sample.

x	$P_{(x)}$	$f = nP_{(x)}$
0	$P_{(x=0)} = \left(1+\dfrac{\bar{x}}{k}\right)^{-k} = 0\cdot0413$	3·31
1	$P_{(x=1)} = \left(\dfrac{k}{1}\right)\left(\dfrac{\bar{x}}{\bar{x}+k}\right) P_{(x=0)} = 0\cdot0851$	6·81
2	$P_{(x=2)} = \left(\dfrac{k+1}{2}\right)\left(\dfrac{\bar{x}}{\bar{x}+k}\right) P_{(x=1)} = 0\cdot1136$	9·09
3	$P_{(x=3)} = \left(\dfrac{k+2}{3}\right)\left(\dfrac{\bar{x}}{\bar{x}+k}\right) P_{(x=2)} = 0\cdot1243$	9·94
4	$P_{(x=4)} = \left(\dfrac{k+3}{4}\right)\left(\dfrac{\bar{x}}{\bar{x}+k}\right) P_{(x=3)} = 0\cdot1211$	9·69
5	$P_{(x=5)} = \left(\dfrac{k+4}{5}\right)\left(\dfrac{\bar{x}}{\bar{x}+k}\right) P_{(x=4)} = 0\cdot1092$	8·74
6	$P_{(x=6)} = \left(\dfrac{k+5}{6}\right)\left(\dfrac{\bar{x}}{\bar{x}+k}\right) P_{(x=5)} = 0\cdot0932$	7·46
7	$P_{(x=7)} = \left(\dfrac{k+6}{7}\right)\left(\dfrac{\bar{x}}{\bar{x}+k}\right) P_{(x=6)} = 0\cdot0764$	6·11
8	$P_{(x=8)} = \left(\dfrac{k+7}{8}\right)\left(\dfrac{\bar{x}}{\bar{x}+k}\right) P_{(x=7)} = 0\cdot0606$	4·85
9	$P_{(x=9)} = \left(\dfrac{k+8}{9}\right)\left(\dfrac{\bar{x}}{\bar{x}+k}\right) P_{(x=8)} = 0\cdot0469$	3·75
10	$P_{(x=10)} = \left(\dfrac{k+9}{10}\right)\left(\dfrac{\bar{x}}{\bar{x}+k}\right) P_{(x=9)} = 0\cdot0355$	2·84

3.2 FOUR WELL-KNOWN FREQUENCY DISTRIBUTIONS

11	$P_{(x=11)} = \left(\dfrac{k+10}{11}\right)\left(\dfrac{\bar{x}}{\bar{x}+k}\right)$ $P_{(x=10)} = 0\cdot0264$		2·11
12	$P_{(x=12)} = \left(\dfrac{k+11}{12}\right)\left(\dfrac{\bar{x}}{\bar{x}+k}\right)$ $P_{(x=11)} = 0\cdot0194$		1·55
13	$P_{(x=13)} = \left(\dfrac{k+12}{13}\right)\left(\dfrac{\bar{x}}{\bar{x}+k}\right)$ $P_{(x=12)} = 0\cdot0140$		1·12
14	$P_{(x=14)} = \left(\dfrac{k+13}{14}\right)\left(\dfrac{\bar{x}}{\bar{x}+k}\right)$ $P_{(x=13)} = 0\cdot0100$		0·80
15	$P_{(x=15)} = \left(\dfrac{k+14}{15}\right)\left(\dfrac{\bar{x}}{\bar{x}+k}\right)$ $P_{(x=14)} = 0\cdot0071$		0·57
16	$P_{(x=16)} = \left(\dfrac{k+15}{16}\right)\left(\dfrac{\bar{x}}{\bar{x}+k}\right)$ $P_{(x=15)} = 0\cdot0050$		0·40
17	$P_{(x=17)} = \left(\dfrac{k+16}{17}\right)\left(\dfrac{\bar{x}}{\bar{x}+k}\right)$ $P_{(x=16)} = 0\cdot0033$		0·26
18	$P_{(x=18)} = \left(\dfrac{k+17}{18}\right)\left(\dfrac{\bar{x}}{\bar{x}+k}\right)$ $P_{(x=17)} = 0\cdot0023$		0·18
19	$P_{(x=19)} = \left(\dfrac{k+18}{19}\right)\left(\dfrac{\bar{x}}{\bar{x}+k}\right)$ $P_{(x=18)} = 0\cdot0016$		0·13
20	$P_{(x=20)} = \left(\dfrac{k+19}{20}\right)\left(\dfrac{\bar{x}}{\bar{x}+k}\right)$ $P_{(x=19)} = 0\cdot0011$		0·09
21	$P_{(x=21)} = \left(\dfrac{k+20}{21}\right)\left(\dfrac{\bar{x}}{\bar{x}+k}\right)$ $P_{(x=20)} = 0\cdot0008$		0·06
22	$P_{(x=22)} = \left(\dfrac{k=21}{22}\right)\left(\dfrac{\bar{x}}{\bar{x}+k}\right)$ $P_{(x=21)} = 0\cdot0005$		0·04
23	$P_{(x=23)} = \left(\dfrac{k+22}{23}\right)\left(\dfrac{\bar{x}}{\bar{x}+k}\right)$ $P_{(x=22)} = 0\cdot0003$		0·02
24	$P_{(x=24)} = \left(\dfrac{k+23}{24}\right)\left(\dfrac{\bar{x}}{\bar{x}+k}\right)$ $P_{(x=23)} = 0\cdot0002$		0·02
		$\Sigma P_{(x)} = 0\cdot9992$	$\Sigma f = 79\cdot94$

3.2.4 Normal distribution and transformations

If there are equal chances of an event occurring one way or another in the positive binomial ($q = p = 0.5$) and k approaches infinity, then the probability series given by $(q+p)^k$ approaches a smooth, symmetrical, bell-shaped curve. This is essentially the normal distribution, which is the distribution associated with continuous variables, *i.e.* measurements rather than counts. The normal distribution is rarely a suitable model for counts, but is important because a large number of methods are associated with it, *e.g.* t-tests, analysis of variance, correlation coefficient. As these methods are fully described in most standard statistical text-books (*e.g.* David 1953, Bailey 1959, Bishop 1966, and for a detailed account, Snedecor & Cochran 1967), a full account is not included here. The use of these methods involves the following conditions:

(1) The data must follow a normal distribution.
(2) The variance of the sample must be independent of the mean.
(3) The components of the variance should be additive.

The positive binomial distribution is approximately normal if the number of sampling units is large (n > 30) and the variance of the sample is not less than 3 (variance $s^2 = kpq$. Therefore the normal

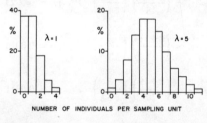

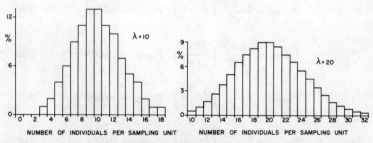

FIGURE 2. Poisson series for various values of λ from 1 to 20. Frequency of each count is expressed as percentage of total count. Compare $\lambda = 10$ with $k = \infty$ in Fig. 4.

3.2 FOUR WELL-KNOWN FREQUENCY DISTRIBUTIONS

approximation can be used when $p = 0.4$ to 0.6 for k between 10 and 30, or when $p = 0.1$ to 0.9 for $k > 30$; and cannot be used when $k < 10$). In the Poisson series, the frequency distribution is very asymmetrical for low values of the parameter λ (estimated by $m = \bar{x} = s^2$), but approaches normality as λ increases in size, and is approximately normal when λ is greater than 10 (see Fig. 2). The

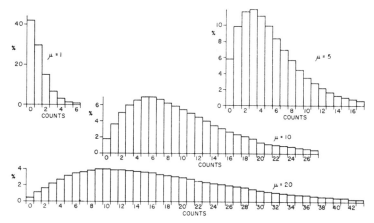

FIGURE 3. Negative binomial frequency distribution for $k = 2.5$ and various values of μ from 1 to 20. Frequency of each count is expressed as percentage of total count.

negative binomial distribution is asymmetrical for a large range of arithmetic means when k is small (*e.g.* $k = 2.5$ in Fig. 3), but approaches normality when k increases and the mean is reasonably large (*e.g.* $\mu = 10$ in Fig. 4). When k approaches infinity, the distribution is identical to that of the Poisson series (*cf.* $\lambda = 10$ in Fig. 2 and $\mu = 10$, $k = \infty$ in Fig. 4). Therefore the first condition of normality can be fulfilled by all three distributions and some methods associated with the normal distribution can then be applied (see section 6).

As the variance and mean tend to increase together in all three distributions (see Taylor's Power Law in section 5.5), the second condition of independent mean and variance is never fulfilled. Therefore some methods, including the *t*-test and analysis of variance, cannot be applied without the risk of considerable errors. This difficulty can be overcome by replacing each count by a suitable mathematical function, *e.g.* the logarithm of the count. Counts are thus *transformed*, and the correct transformation should normalise

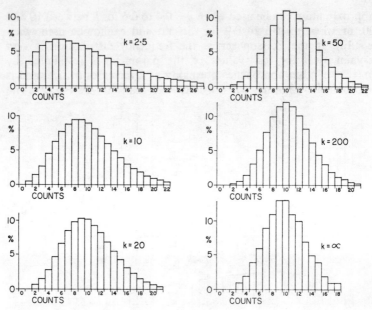

FIGURE 4. Negative binomial frequency distributions for $\mu = 10$ and various values of k from 2·5 to ∞. Frequency of each count is expressed as percentage of total count. Compare $k = \infty$ with $\lambda = 10$ in Fig. 2.

the frequency distribution of the counts, eliminate the dependence of the variance on the mean, and ensure that the components of the variance are additive (for the analysis of variance). The choice of the correct transformation depends upon the original frequency distribution of the counts (see Table 4). If the number of sampling units is too small for the counts to be arranged in a frequency distribution, then the relationship between variance and arithmetic mean can be used to choose a suitable transformation (see Table 4). As the variance of a bottom sample is often greater than the arithmetic mean, $\log(x+1)$ is probably the most useful transformation. Although it can usually be assumed that the variance and mean are not independent before transformation, this can be checked by plotting mean against variance on a log/log scale (see Fig. 6 and section 5.5). Example 7 illustrates the effect of transformations on (a) the frequency distribution of the counts, and (b) the relationship between mean and variance.

Analysis of variance, correlation coefficients, t-tests and other methods associated with the normal distribution (see standard

3.2 FOUR WELL-KNOWN FREQUENCY DISTRIBUTIONS

TABLE 4. TRANSFORMATIONS

A wide range of transformations can also be obtained from Taylor's power law (sections 5.5 and 6.2.4, example 21). x is a particular count. $\sinh^{-1} x$ is the inverse function of the hyperbolic sine ($\sinh x$).

Original distribution	Distribution not known	Transformation	Special conditions
Poisson	$s^2 = \bar{x}$	replace x by $\sqrt{x}$	No counts less than 10
Poisson	$s^2 = \bar{x}$	replace x by $\sqrt{x+0.5}$	Some counts less than 10
Negative binomial		replace x by $\sinh^{-1} \sqrt{\dfrac{x+0.375}{k-2(0.375)}}$	k greater than 5
Negative binomial		replace x by $\log(x+k/2)$	k between 2 and 5
	$s^2 > \bar{x}$	replace x by $\log x$	no zero counts
	$s^2 > \bar{x}$	replace x by $\log(x+1)$	some zero counts

textbooks) are performed on the transformed counts. When the analyses are complete, the arithmetic mean of the transformed counts has to be transformed back to the original scale and thus becomes a *derived mean*; e.g. for a square root transformation, the mean transformed count must be squared; and for a log $(x+1)$ transformation, the antilog of the mean transformed count must be taken and one subtracted. As the derived mean is smaller than the arithmetic mean of the original counts before transformation, it is not comparable with arithmetic means obtained by direct averaging. Therefore small adjustments have to be made to the derived means:

(1) square root transformation—the mean square root is first squared, 0·5 is subtracted if necessary, and then the variance of the transformed counts is added to the derived mean;

(2) log transformation—1·15 times the variance of the transformed counts is first added to the mean transformed count before transforming back to the antilog.

The final value (derived mean + adjustment) is usually in good agreement with means obtained by direct averaging. A more detailed account of transformations is given in Chapter 8 of Quenouille (1950).

Although transformations are somewhat complicated, they are an essential procedure before the application of most methods associated with the normal distribution. An alternative procedure is to use *distribution-free* methods on the original counts (see section 7.2).

Example 7A. Effect of a transformation on a negative binomial distribution

In a large sample of 100 sampling units, $\bar{x} = 5 \cdot 19$, $\hat{k} = 2 \cdot 5$, and the whole frequency distribution is:

f	6	10	12	12	11	10	8	7
x	0	1	2	3	4	5	6	7
$\log(x+k/2)$	0·0969	0·3522	0·5119	0·6284	0·7202	0·7959	0·8603	0·9165

f	6	4	4	3	2	2	1	1	1
x	8	9	10	11	12	13	14	15	16
$\log(x+k/2)$	0·9661	1·0107	1·0512	1·0881	1·1222	1·1538	1·1833	1·2109	1·2368

where $\log(x+k/2)$ is the appropriate transformation.

A comparison of frequency diagrams for the original and transformed counts indicates that the asymmetry of the original distribution is considerably reduced after transformation (Fig. 5).

Example 7B. Effect of a transformation on the relationship between mean and variance

The following counts were obtained in six samples, each of 20 sampling units:

Samples:	1	2	3	4	5	6
	0	3	1	6	7	12
	2	1	5	1	2	7
	1	1	2	5	6	10
	0	1	0	7	9	15
	0	4	2	4	5	9
	1	0	5	1	2	6
	1	1	2	6	7	13
	0	4	1	5	6	11
	1	3	3	3	4	8
	1	3	4	3	3	7
	0	5	1	5	5	10
	2	3	3	3	3	8
	1	2	2	4	6	11
	0	2	4	3	4	8
	0	1	0	8	8	14
	2	1	3	4	5	9
	3	2	4	2	2	6
	0	2	4	2	3	7
	1	2	3	4	4	9
	1	0	6	2	1	5
Mean $\bar{x}$	0·85	2·05	2·75	3·9	4·6	9·25
Variance s^2	0·77	1·84	2·83	3·67	4·78	7·57

3.2 FOUR WELL-KNOWN FREQUENCY DISTRIBUTIONS

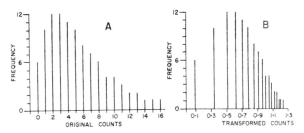

FIGURE 5. Frequency diagrams for original counts (A) and transformed counts (B) of a negative binomial frequency distribution. $\bar{x} = 5.19$, $\hat{k} = 2.5$, $n = 100$.

When mean is plotted against variance on a log/log scale, the two statistics tend to increase together (see Fig. 6) and therefore variance and mean are not independent. As the mean and variance are approximately equal for each sample, the square root transformation is most suitable (x replaced by $\sqrt{x+0.5}$ as some counts less than 10). The following means and variances were obtained for the transformed counts:

Samples	1	2	3	4	5	6
Mean x	1·10	1·54	1·73	2·05	2·21	3·09
Variance s^2	0·14	0·19	0·27	0·21	0·24	0·20

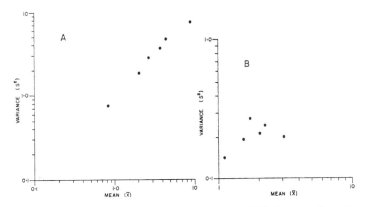

FIGURE 6. Relationship between mean and variance, (A) before transformation. (B) after a square root transformation. Values are plotted on a log/log scale.

There is no longer a tendency for mean and variance to increase together (see Fig. 6), and therefore the condition of independence is now fulfilled.

3.2.5 *The binomial family*

The relationships between the various members of the Binomial family are summarised below:

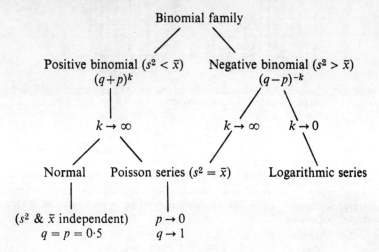

IV THE SPATIAL DISPERSION OF A POPULATION

The *spatial dispersion of a Population* describes the spatial distribution of the individuals in the Population (ecological meaning, see section 2.1). Temporal changes in dispersion will usually occur, and different stages of the same species will often show different patterns of dispersion. As it affects the analysis of samples (sections 3, 6, 7) and the sampling programme (section 8), the dispersion of a Population is also of practical importance.

The individuals of a Population can follow three basic types of spatial distribution:

(1) a *random* distribution,

(2) a *regular* distribution (= under-dispersion, or uniform distribution, or even distribution),

(3) a *contagious* distribution (= over-dispersion, or clumped distribution, or aggregated distribution).

Although the terms regular and contagious can be criticised, they are most frequently used and cause less confusion than alternative terms (given in parentheses). The three basic distributions are shown in Fig. 7, and it must be remembered that the contagious distribution can show many other patterns (see section 5). It must also be remembered that the three types of distribution can overlap, *e.g.* a contagious distribution can result from randomly-distributed groups with regularly-distributed individuals in each group. This overlap of distributions will produce different patterns, and the detection of a contagious distribution will depend upon the scale of the pattern relative to the size of the sampling unit (see section 5.4).

The dispersion of a Population determines the relationships between the variance (σ^2) and the arithmetic mean (μ) thus:

(1) random distribution—variance equal to mean ($\sigma^2 = \mu$),

(2) regular distribution—variance less than mean ($\sigma^2 < \mu$),

(3) contagious distribution—variance greater than mean ($\sigma^2 > \mu$).

Known mathematical distributions are suitable models for the three possible relationships between variance and mean (see section 3.2), and these models can also be applied to the corresponding patterns of dispersion. A Poisson series ($\sigma^2 = \mu$) is a suitable model for a random distribution, and the positive binomial ($\sigma^2 < \mu$) is an approximate model for a regular distribution. The negative binomial ($\sigma^2 > \mu$) is often used for contagious distributions, but is only one of several possible models (see section 5).

4.1 Random distribution

A random distribution is usually the first hypothesis to be considered. In a random distribution, there is an equal chance of an individual occupying any point in an area of bottom and the presence of an individual does not influence the position of a nearby individual. The striking feature of a random distribution is the lack of any system; *e.g.* some individuals occur in groups and others are equally spaced, some individuals are very close together and others are wide apart (see Fig. 7A).

Agreement with a Poisson series is the accepted test for randomness (sections 4.1.1, 4.1.2, 4.1.3), and the use of a Poisson series as

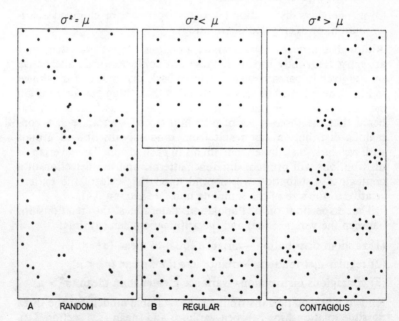

FIGURE 7. The three types of spatial distribution. (A) Random, (B) Regular (upper-ideal form, lower-normal form), (C) Contagious (see also Fig. 9).

a mathematical model requires four definite conditions (page 22). No tests can prove randomness, and agreement with a Poisson series simply means that the hypothesis of randomness is not disproved. Although the hypothesis is usually accepted after agreement with a Poisson series, there is always a possibility that non-randomness exists and cannot be detected. Therefore it is important to consider the possible reasons for a random distribution before the hypothesis is accepted. A random distribution could be due to either: (1) the influence of a single factor whose values are themselves randomly distributed, or (2) chance effects.

Important environmental factors rarely follow a random distribution and therefore the first explanation can rarely be considered. One possible exception is oviposition behaviour, which could produce a random distribution of first instars if the eggs were laid singly and at random. This explanation cannot be applied to many species which lay their eggs in clumps.

If the first explanation is rejected, the second must be considered together with the logical conclusions that: (1) environmental factors have no effect, or a relatively small effect, on the dispersion of the Population; (2) there is no tendency for individuals in the Population to avoid or to move towards each other.

These conclusions may be difficult to accept, and therefore it can only be concluded that non-randomness is present but cannot be detected by sampling techniques in the field.

The problems of detecting non-randomness will be discussed later (section 5.4), but the important effect of quadrat size must be briefly mentioned. If the size of the sampling unit (*i.e.* quadrat size) is much larger or much smaller than the average size of clumps of individuals, and these clumps are regularly or randomly distributed; then the dispersion of the Population is apparently random and the non-randomness is not detected. Most samplers will detect non-randomness if the sampling unit is small (about $0·05$ m^2 or less), but even a small quadrat will not detect a contagious distribution if there are only a few individuals in each clump. Therefore the dispersion of a Population is effectively random if the density of the Population is low. The tendency to randomness often increases with the age of a Population, but this change could be due to a decrease in Population density or to the division of larger clumps into several smaller clumps.

Therefore a random distribution is usually a suitable hypothesis for low density Populations, but the implications of randomness must be carefully considered before the hypothesis is accepted for other Populations. Occam's razor (see Oxford English Dictionary) requires that, unless we can reject the hypothesis of a random distribution, we use the simplest possible hypothesis; and there are many situations in which it is not really profitable to do anything else! The practical advantage of a random distribution is the use of a Poisson series with its relatively simple methods for calculating confidence limits (section 6.2.2) and comparing samples (section 7.1.2).

4.1.1 *A rapid test for agreement with a Poisson series*

Confidence limits for a Poisson variable (Fig. 19) provide an approximate test for agreement with a Poisson series. If all the counts of a sample lie within the confidence limits for the sample mean, then the hypothesis of randomness is not disproved. For example, the counts of example 1 range from 7 to 15, the mean = 11·27, and the 95% confidence limits for the mean are about 5·5 to 20 (from Fig. 19). As all the counts lie within these confidence limits, the counts could come from a Poisson series and therefore the dispersion of the parent Population could be random. The disadvantages of this quick test are: (1) the result will be the same for random and regular distributions, and (2) the dispersion of the parent Population could still be random if one or two counts lie outside the confidence limits. Therefore this test is often unreliable, and should always be checked by either, or both, of the following tests.

4.1.2 *Variance to mean ratio*

This test is based on the equality of variance and mean in a Poisson series. The variance to mean ratio, or *index of dispersion* (I), will approximate to unity if there is agreement with a Poisson series.

$$I = \frac{\text{sample variance}}{\text{theoretical variance}} = \frac{s^2}{\bar{x}} = \frac{\Sigma(x-\bar{x})^2}{\bar{x}(n-1)}$$

where s^2 = variance, $\bar{x}$ = arithmetic mean, and n = number of sampling units. This index of dispersion will often depart from unity, and the significance of these departures from unity is assessed by reference to a table of χ^2 (chi squared). The expression $I(n-1)$ gives a good approximation to χ^2 with $n-1$ degrees of freedom and therefore:

$$\chi^2 = I(n-1) = \frac{s^2(n-1)}{\bar{x}} = \frac{\Sigma(x-\bar{x})^2(n-1)}{\bar{x}(n-1)} = \frac{\Sigma(x-\bar{x})^2}{\bar{x}}$$

4.1 RANDOM DISTRIBUTION

The χ^2 distribution has many uses, and percentage points of the distribution are given in Fig. 8 or more accurately in Pearson & Hartley's (1966) Table 8. Agreement with a Poisson series is accepted at the 95% probability level ($P > 0.05$) if the χ^2 value lies between the appropriate 5% significance levels for $n-1$ degrees of freedom. If agreement is perfect, $I = 1$ and $\chi^2 = n-1$. The significance levels are given in Fig. 8, and in the columns headed $Q = 0.975$ and $Q = 0.025$ for v degrees of freedom ($v = n-1$) in Pearson & Hartley's (1966) Table 8.

If the sample is large (n > 31), it is assumed that $\sqrt{2\chi^2}$ is distributed normally about $\sqrt{2v-1}$ with unit variance. Agreement with a Poisson series is then accepted ($P > 0.05$) if the absolute value of d (*i.e.* without regard of sign) is less than 1·96 in:

$$d = \sqrt{2\chi^2} - \sqrt{2v-1}$$

where d is a normal variable with zero mean and unit standard deviation; and v is the number of degrees of freedom.

Departures from a Poisson series, and hence a random distribution, can occur in two directions. The χ^2 value can be:

(1) less than expected (χ^2 < expected value for $Q = 0.975$, or $d > 1.96$ with a negative sign), and therefore a regular distribution is suspected ($s^2 < \bar{x}$); or

(2) greater than expected (χ^2 > expected value for $Q = 0.025$, or $d > 1.96$ with a positive sign), and therefore a contagious distribution is suspected ($s^2 > \bar{x}$).

If the sample is large, the result of this test should always be checked by the χ^2 test of goodness-of-fit.

Example 8. χ^2 *test (variance to mean ratio) for agreement with a Poisson series; small samples* (n < 31)

(8A) The counts of *Baëtis rhodani* in examples 1 and 2 were obtained from a random sample of 11 sampling units. Arithmetic mean of the sample ($\bar{x}$) = 11·273, variance (s^2) = 7·415, number of sampling units (n) = 11 and therefore degrees of freedom = $n - 1 = 10$. There are two methods of calculating χ^2:

$$\chi^2 = \frac{s^2(n-1)}{\bar{x}} = \frac{7\cdot415(10)}{11\cdot273} = 6\cdot578$$

or

$$\chi^2 = \frac{\Sigma(x-\bar{x})^2}{\bar{x}} = \frac{\Sigma(x)^2 - \bar{x}\Sigma x}{\bar{x}} = \frac{1472 - 11\cdot273(124)}{11\cdot273}$$

$$= \frac{74\cdot15}{11\cdot273} = 6\cdot578$$

42 THE SPATIAL DISPERSION OF A POPULATION

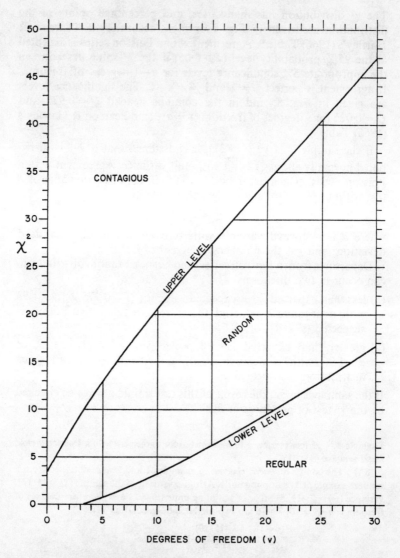

FIGURE 8. The 5% significance levels of χ^2. If χ^2 value between significance levels, agreement with Poisson series is accepted at 95% probability level ($P > 0.05$).

EXAMPLE: $\chi^2 = 6.578$ and degrees of freedom (v) = 10 (see example 8A). First locate degrees of freedom and draw vertical line. χ^2 value clearly lies between upper and lower levels of 3 and 20.5. Therefore agreement with Poisson accepted ($P > 0.05$). χ^2 below lower level indicates regular distribution. χ^2 above upper level indicates contagious distribution.

4.1 RANDOM DISTRIBUTION

This χ^2 value clearly lies between the 5% significance levels in Fig. 8, and between the values of 3·247 (Q = 0·975) and 20·483 (Q = 0·025) for 10 degrees of freedom (v = 10) in Pearson & Hartley's (1966) Table 8. Therefore agreement with a Poisson series is accepted at the 95% probability level ($P > 0·05$), and the hypothesis of randomness is not disproved. It is therefore possible that the nymphs of *B. rhodani* are randomly distributed on the bottom of the stream, *i.e.* the dispersion of the Population is random.

(8B) The following counts of *B. rhodani* were obtained from a random sample of 5 sampling units; 98, 22, 72, 214, 67. $\bar{x} = 94·60$, $s^2 = 5202·80$, $n = 5$ and therefore degrees of freedom = 4.

$$\chi^2 = \frac{s^2(n-1)}{\bar{x}} = \frac{5202·80(4)}{94·60} = 219·99$$

This χ^2 value lies well above the upper 5% significance level in Fig. 8, and well above the value of 11·143 (Q = 0·025) for 4 degrees of freedom (v = 4) in Pearson & Hartley's (1966) Table 8. Therefore agreement with a Poisson series is rejected at the 95% probability level ($P < 0·05$) and also at the 99% probability level ($P < 0·01$ and $Q = 0·005$ in Table 8). The high value of χ^2 indicates that the nymphs of *B. rhodani* are definitely clumped on the bottom of the stream, *i.e.* the dispersion of the Population is contagious.

Both these samples of nymphs of *B. rhodani* were taken from the same section of stream with a shovel sampler, but in different months. The large difference between the χ^2 values clearly shows that the dispersion pattern of the same species can change markedly though the year. In the second example, the contagious distribution was due to clumps of tiny nymphs which were abundant at the time of sampling.

Example 9. χ^2 **test (variance to mean ratio) for agreement with a Poisson series; large samples (n > 31)**

(9A) The counts of *B. rhodani* in example 3 were obtained from a large random sample of 80 sampling units. $\bar{x} = 10·1250$, $s^2 = 8·5918$, and degrees of freedom $(v) = n - 1 = 79$.

$$\chi^2 = \frac{s^2(n-1)}{\bar{x}} = \frac{8·5918(79)}{10·1250} = 67·0373$$

As the sample is large, the normal variable (d) is calculated thus:

$$d = \sqrt{2\chi^2} - \sqrt{2v-1} = \sqrt{134·0746} - \sqrt{157}$$
$$= 11·580 - 12·530$$
$$= -0·950$$

As this value of d is less than 1·96, agreement with a Poisson series is accepted at the 95% probability level ($P > 0·05$) and the hypothesis of randomness is not disproved. This result is checked in example 10 by the χ^2 test of goodness of fit.

(9B) The counts of *Gammarus pulex* in example 6 were obtained from a large random sample of 80 sampling units.

$\bar{x} = 5\cdot3125$, $\qquad s^2 = 13\cdot534$, $\qquad$ and degrees of freedom $(v) = 79$.

$$\chi^2 = \frac{s^2(n-1)}{\bar{x}} = \frac{13\cdot534(79)}{5\cdot3125} = 201\cdot2585$$

As the sample is large, the normal variable (d) is calculated thus:

$$d = \sqrt{2\chi^2} - \sqrt{2v-1} = \sqrt{402\cdot5170} - \sqrt{157} = +7\cdot532$$

As this value of d is greater than 1·96, agreement with a Poisson series is rejected at the 95% probability level and also at the 99% probability level ($d = 2\cdot58$). The high value of d with a positive sign is a strong indication that the dispersion of the Population is contagious ($\sigma^2 > \mu$).

4.1.3 χ^2 test for "goodness-of-fit"

If a sample is large enough for the counts to be arranged in a frequency distribution (see example 3), then the observed frequency distribution of the counts can be compared with the expected frequency distribution from a mathematical model. The model is a good fit to the original counts when the observed and expected frequencies agree. This "goodness-of-fit" is tested by χ^2 and

$$\chi^2 = \sum \frac{(\text{observed} - \text{expected})^2}{\text{expected}}$$

The χ^2 value is first calculated for the observed and expected values in each frequency class, and the total χ^2 for the whole frequency distribution is then referred to tables of χ^2. It is usually recommended that some frequencies are combined so that no expected values are less than 5. Cochran (1954, see also Snedecor & Cochran 1967) considers that this restriction weakens the sensitivity of the test, and he suggests that no expected values should be less than 1. The number of degrees of freedom (v) is given by:

v = (number of frequency classes after any combinations) −
  − (number of estimated parameters) − 1

In a Poisson series, only one parameter (λ) is estimated from the data and therefore v = (number of frequency classes) − 2.

Agreement with the model (*e.g.* a Poisson series) is accepted at the 95% probability level ($P > 0\cdot05$) if the χ^2 value is less than the 5% point for χ^2 with v degrees of freedom. The 5% significance level is slightly less than the upper level in Fig. 8, and is given in the column headed $Q = 0\cdot050$ in Pearson & Hartley's (1966) Table 8. If v is greater than 30, agreement with the model is accepted ($P > 0\cdot05$) when the absolute value of d is less than 1·645 in :

$$d = \sqrt{2\chi^2} - \sqrt{2v-1}$$

4.1 RANDOM DISTRIBUTION

Example 10. χ^2 **test (goodness-of-fit) for agreement with a Poisson series.**

The counts of *B. rhodani* in example 3 have already been tested for agreement with a Poisson series by the variance to mean ratio (example 9A). The 80 counts were arranged in a frequency distribution in example 3 and these observed frequencies are given in Table 5. The arithmetic mean of the sample $(\bar{x})$ = 10·1250, and therefore the estimate (m) of the Poisson parameter = 10·1250. The expected frequency distribution for a Poisson series $(m = 10·125)$ was calculated in example 5, and these expected frequencies are given in Table 5. As some expected frequencies were less than 1 (see Table 2), the values in both tails of the distribution were combined (see Table 5). Therefore there are 15 frequency classes and the number of degrees of freedom $(v) = 15 - 2 = 13$.

TABLE 5. χ^2 TEST FOR GOODNESS-OF-FIT OF A POISSON DISTRIBUTION

to a sample with $m = 10·125$. x is a particular count in each frequency class, Obs. is the observed frequency of that particular count, Exp. is the expected frequency of the count, and χ^2 is given for each frequency class.

x	Obs.	Exp.	Obs. – Exp.	χ^2
0 – 4	2	2·16	—0·16	0·01
5	2	2·84	—0·84	0·25
6	4	4·80	—0·80	0·13
7	7	6·94	0·06	0·00
8	10	8·78	1·22	0·17
9	10	9·88	0·12	0·00
10	10	10·00	0	0·00
11	10	9·21	0·79	0·07
12	8	7·77	0·23	0·01
13	6	6·05	—0·05	0·00
14	4	4·38	—0·38	0·03
15	4	2·95	1·05	0·37
16	2	1·87	0·13	0·01
17	1	1·11	—0·11	0·01
18 or more	0	1·28	—1·28	1·28
Total	80	80·02		2·34

χ^2 values for each frequency class are given in the fifth column of Table 5 and the total χ^2 value = 2·34. This χ^2 value is well below the 5% point of 22·36 $(Q = 0·050$ and $v = 13)$ in Pearson & Hartley's (1966) Table 8. Therefore the model is a good fit to the original counts and agreement with a Poisson series is accepted at the 95% probability level $(P > 0·05)$.

4.2 REGULAR DISTRIBUTION

The dispersion of a Population is regular when the individuals in the Population are relatively crowded and move away from each other. Under these conditions, the number of individuals per sampling unit approaches the maximum possible, the variance of the population is less than the mean ($\sigma^2 < \mu$), and the positive binomial is an approximate mathematical model. The characteristic feature of a regular distribution is the uniform spacing of the individuals in the Population (lower Fig. 7B), and in a perfectly regular distribution the individuals are equidistant from each other (upper Fig. 7B).

Territorial behaviour will often produce a uniform spacing of individuals, and therefore the dispersion of sedentary invertebrates may be regular over a small area of bottom, *e.g.* tube-dwelling larvae of Chironomidae, net-spinning larvae of Trichoptera, larvae of Simuliidae. Although the dispersion of most species is not regular over a large area of bottom, it is sometimes regular within clumps of individuals. For example, nymphs of *Baetis rhodani* may show a contagious distribution over a large area of bottom and a regular distribution within a dense group of nymphs on the upper surface of a stone at night. The nymphs feed on algae, and the regular distribution is presumably due to each nymph having a definite grazing territory.

Therefore a regular distribution will rarely describe the dispersion of a Population over a large area, but will sometimes describe the dispersion in a small area. The study of micro-dispersion requires special techniques, *e.g.* nearest neighbour methods (Southwood 1966, p. 40), and a description of these techniques is beyond the scope of this account. If the animals are visible, their distribution can be photographed (*e.g.* Heywood & Edwards 1961) or marked on the transparent bottom of a viewing box (see example 11). Edgar & Meadows (1969) used nearest-neighbour methods to investigate the spatial distribution of larvae of *Chironomus riparius*, and give a detailed account of their statistical methods.

4.2.1 *The use of a positive binomial distribution as an approximate model for a regular distribution*

The expected frequency distribution of a positive binomial is given by $n(q+p)^k$; where n = number of sampling units, p = probability of any point in the sampling unit being occupied by an individual, $q = 1-p$, and k = maximum possible number of individuals a sampling unit could contain (see section 3.2.1). Estimates

4.2 REGULAR DISTRIBUTION

of the parameters k, p, and q are obtained from the sample. The highest count in the sample is a rough estimate of k, but may be too low. A more accurate estimate is obtained from the mean and variance thus:

$$\text{estimate of } k = \hat{k} = \frac{\bar{x}^2}{\bar{x} - s^2}$$

to nearest whole number (k must be an integer).
Estimates of p and q are:

$$\hat{p} = \frac{\bar{x}}{\hat{k}} \quad \text{and} \quad \hat{q} = 1 - \hat{p}$$

Example 11. Test for agreement with a positive binomial

A group of *Simulium* larvae on a large stone was observed through a viewing box with a grid marked on the bottom. 20 squares (each 2 by 2 cm) were randomly selected from a total of 63 and the larvae in each square were counted. The counts are summarised in a frequency distribution:

$$\begin{array}{ccccc} f & 0 & 5 & 10 & 5 \\ x & 1 & 2 & 3 & 4 \end{array}$$

where x is a particular count and f is the frequency of that particular count in the sample.

$$\text{Number of sampling units } (n) = \Sigma f = 20$$

$$\text{Arithmetic mean of sample } (\bar{x}) = \frac{\Sigma fx}{\Sigma f} = \frac{60}{20} = 3$$

$$\text{Variance of sample } (s^2) = \frac{\Sigma(fx^2) - \bar{x}\Sigma fx}{n-1}$$

$$= \frac{190 - 180}{19}$$

$$= 0.5263$$

$$\chi^2 \text{ (variance to mean ratio)} = \frac{s^2(n-1)}{\bar{x}} = \frac{0.5263(19)}{3} = 3.3332$$

This χ^2 value lies well below the lower 5% significance level in Fig. 8, and well below the value of 8·907 ($Q = 0.975$) for 19 degrees of freedom ($\nu = 19$) in Pearson & Hartley's (1966) Table 8. Therefore agreement with Poisson series is rejected ($P < 0.05$ and also $P < 0.01$), and the spatial distribution of the larvae is regular.

This conclusion can be checked by a χ^2 test for goodness-of-fit of a positive binomial distribution to the sample.
Estimates of k, p, q are:

$$\hat{k} = \frac{\bar{x}^2}{\bar{x} - s^2} = \frac{9}{2.4737} = 3.6383 \approx 4 \text{ (to nearest integer)}$$

$$\hat{p} = \frac{\bar{x}}{\hat{k}} = \frac{3}{4} = 0.75, \quad \hat{q} = 1 - \hat{p} = 0.25$$

The expected frequencies were calculated in example 4 and are compared with the observed frequencies in Table 6. There are four frequency classes after combinations, and the parameters p and k were estimated from the sample (q was estimated from $\hat{p}$). Therefore the number of degrees of freedom (v) = $4-3 = 1$.

The total χ^2 value of 1·73 is below the 5% point of 3·84 ($Q = 0.050$ and $v = 1$) in Pearson & Hartley's (1966) Table 8. Therefore the positive binomial is a good fit to the original counts. As the maximum number of larvae per square (4 cm^2) is 4 and the distribution of the larvae is regular, it is concluded that the territory of each larva is at least 1 cm^2.

TABLE 6. χ^2 TEST FOR GOODNESS-OF-FIT OF A POSITIVE BINOMIAL DISTRIBUTION

to a sample. x, Obs., and Exp. are defined in Table 5.

x	Obs.	Exp.	Obs.−Exp.	χ^2
0 } 1	0	1·02	−1·02	1·02
2	5	4·22	0·78	0·14
3	10	8·44	1·56	0·29
4	5	6·33	−1·33	0·28
Total	20	20·00		1·73

4.3 SUMMARY

The dispersion of a Population describes the spatial distribution of individuals in the Population, and can be random ($\sigma^2 = \mu$), regular ($\sigma^2 < \mu$), or contagious ($\sigma^2 > \mu$).

Random (section 4.1).

Agreement with a Poisson series is the accepted test for a random distribution and there are three possible tests:

(1) Confidence limits (section 4.1.1)—quick test, often unreliable.

(2) Variance to mean ratio (section 4.1.2).

$$\chi^2 = \frac{s^2(n-1)}{\bar{x}}$$

Agreement with Poisson if χ^2 between 5% significance levels in Fig. 8 or Table 8 in Pearson & Hartley (1966). Degrees of freedom (v) = $n-1$.

(A) Small samples ($n < 31$):

If $\chi^2 <$ expected value for $Q = 0.975$, then dispersion regular ($s^2 < \bar{x}$).

If $\chi^2 >$ expected value for $Q = 0.025$, then dispersion contagious ($s^2 > \bar{x}$).

4.3 SUMMARY

(B) Large samples ($n > 31$): substitute χ^2 and v degrees of freedom in

$$d = \sqrt{2\chi^2} - \sqrt{2v-1}$$

If $d > 1.96$ with negative sign, then dispersion regular.
If $d > 1.96$ with positive sign, then dispersion contagious.

(3) χ^2 test for goodness-of-fit (section 4.1.3)—only large samples. Observed frequency distribution from sample is compared with expected frequency distribution from model (*e.g.* Poisson series).

$$\chi^2 = \sum \frac{(\text{Observed} - \text{Expected})^2}{\text{Expected}}$$

Refer χ^2 to Table 8 in Pearson & Hartley (1966). Agreement with Poisson is accepted ($P > 0.05$) if χ^2 below 5% significance point ($Q = 0.050$) for v degrees of freedom (v = number of frequency classes -2).

If agreement with Poisson accepted, *consider carefully* the implications of randomness before accepting that dispersion of population is random (section 4.1).

Regular (section 4.2).

A regular distribution will rarely describe the dispersion of a Population over a large area, but will sometimes describe the dispersion of sedentary species in a small area and the distribution of individuals within clumps. The positive binomial distribution is an approximate model for a regular distribution (section 4.2.1).

V CONTAGIOUS DISTRIBUTIONS

5.1 THE DIVERSITY OF CONTAGIOUS DISTRIBUTIONS

The spatial dispersion of a Population (ecological meaning, see section 2.1) is seldom random or regular, but is frequently contagious, with the variance significantly greater than the mean (see χ^2 test, section 4.1.2). There are always definite clumps or patches of individuals in a contagious distribution, but the final pattern varies considerably (Figs. 7C, 9). The frequent occurrence of contagious distributions is not surprising, for there are many environmental factors which are unevenly distributed. There is also a tendency for some species to aggregate and thus produce a contagious distribution without the influence of environmental factors. The final dispersion pattern depends upon the size of the clumps, the distance between clumps, the spatial distribution of the clumps, and the spatial distribution of individuals within clumps. One common pattern is

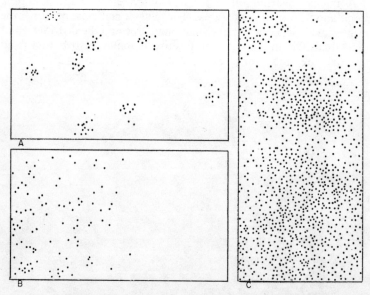

FIGURE 9. Different types of contagious distribution. (A) Small clumps. (B) Large clumps with individuals randomly distributed in each clump. (C) Large clumps with individuals uniformly distributed in each clump.

5.2 NEGATIVE BINOMIAL DISTRIBUTION

due to patches of high density (clumps) on a general background of low density (Fig. 7C). Different species will usually show different contagious distributions within the same habitat, and the dispersion pattern of one species may vary within a small area of bottom. Therefore there are diverse patterns of contagious distributions and it is difficult to find a mathematical distribution which fits all these patterns. Several mathematical models have been proposed, and the negative binomial distribution is probably the most useful of these models. Although Taylor's power law is not strictly a distribution, it is another useful model which can be applied to a wide range of dispersion patterns (section 5.5).

5.2 Negative binomial distribution

This distribution has two parameters (arithmetic mean μ and exponent k) and was fully described in section 3.2.3. The distribution has only one mode (= most frequent count), and therefore additional peaks in the observed distributions are attributed to random sampling. As the negative binomial can be derived from several mathematical and biological models, it can be applied to a wide diversity of contagious distributions.

The following models are appropriate to bottom samples, and the notes may help to explain the models:

(1) *True contagion.* The presence of one individual or event increases the chance that another will occur in the same unit. Consideration of this model requires some knowledge of the behaviour of a species, *e.g.* an insect may lay its eggs in masses and thus produce a contagious distribution.

(2) *Constant birth-death-immigration rates.* The growth of a Population with constant rates of birth and death per individual and of immigration per unit of time leads to a negative binomial for Population size. This model is a stochastic process which could lead to model 3.

(3) *Randomly distributed clumps.* If clumps of individuals are distributed at random over the bottom and the numbers of individuals in the clumps are distributed independently in a logarithmic distribution, then a negative binomial will result. This model fits some counts of bacteria (Quenouille 1949, and statistical note in Jones,

Mollison & Quenouille 1948). The mean number of clumps per sampling unit (m_1) and the mean number of individuals per clump (m_2) are given by:

$$m_1 = k \log_e \left(1 + \frac{\mu}{k}\right) \quad \text{and} \quad m_2 = \frac{\mu}{m_1} = \frac{\mu/k}{\log_e(1 + \mu/k)}$$

where $m_1 \cdot m_2 = \mu$ (Anscombe 1950). The size of the sampling unit (quadrat size) affects the values of μ and m_1, but not m_2 which should remain constant over a short period of time. When m_2 is constant, the ratio μ/k is also constant. Therefore k is always directly proportional to μ, and quadrat size must affect the values of k and μ in the same way.

(4) *Heterogeneous Poisson sampling (compound Poisson distribution)*. The Poisson parameter (λ) varies randomly from occasion to occasion and has a Pearson Type III distribution, proportional to a χ^2 distribution with $2k$ degrees of freedom. Therefore $\lambda \, 2k/\mu$ also has a χ^2 distribution (Arbous & Kerrich 1951), and the mean value of $\lambda = \mu \chi^2 / 2k$, where χ^2 is found in Pearson & Hartley's (1966) Table 8 for $Q = 0.5$ and $v = 2k$. Southwood (1966, p. 34) equates the mean value of λ with the mean number of individuals per clump (m_2), and therefore the mean number of clumps per sampling unit is

$$m_1 = \frac{\mu}{m_2} = \frac{2k}{\chi^2}$$

These formulae imply that m_2 is always slightly less than μ, that m_1 is close to unity for most values of k, and that m_1 is not greatly influenced by quadrat size or k. As these conclusions are difficult to accept, the formulae for m_1 and m_2 are of doubtful value.

These four models are often quoted; but there are probably other, as yet untried, hypotheses which are applicable to samples from a negative binomial. As it is usually difficult to choose the appropriate model for a sample, the fundamental models are of limited value until more is known about the implications and application of each model. The four models do show, however, that divergent hypotheses can lead to the same negative binomial distribution. Bliss & Fisher (1953) give an example where the same frequency distribution in accident statistics could have arisen from two distinct and contradictory hypotheses! Therefore the negative binomial is often a good empirical description of a distribution, but agreement with a negative binomial should not be used as the sole basis for justifying one particular hypothesis.

5.2.1 Tests for agreement with a negative binomial; large samples (n > 50)

As the counts of a large sample can always be arranged in a frequency distribution, the χ^2 test for "goodness-of-fit" (section 4.1.3) is the simplest test for agreement with a negative binomial. The maximum-likelihood method gives the most accurate estimate of the parameter k (section 3.2.3), and should always be used with large samples. The second parameter μ is always estimated by the arithmetic mean ($\bar{x}$) of the sample. As two parameters (μ and k) are estimated from the sample, the number of degrees of freedom $(\nu) = $ (number of frequency classes after any combinations) $- 3$.

Example 12. χ^2 **test (goodness-of-fit) for agreement with a negative binomial distribution**

A shovel sampler was used to take a large random sample of 80 sampling units from the bottom of a stony stream, and freshwater shrimps (*Gammarus pulex*) were counted in each sampling unit (0·05m²). The counts were arranged in a frequency distribution in example 6 and these observed frequencies are given in Table 7. The variance ($s^2 = 13\cdot534$) of the sample is significantly greater than the arithmetic mean ($\bar{x} = 5\cdot3125$), and therefore the dispersion of the Population is contagious (see χ^2 test, example 9B). An estimate of k ($\hat{k} = 3\cdot3588$) and the expected frequencies of the negative binomial distribution were calculated in

TABLE 7. χ^2 TEST FOR GOODNESS-OF-FIT OF A NEGATIVE BINOMIAL DISTRIBUTION

to a sample. x, Obs., and Exp. are defined in Table 5.

x	Obs.	Exp.	Obs. − Exp.	χ^2
0	3	3·31	−0·31	0·03
1	7	6·81	0·19	0·01
2	9	9·09	−0·09	0·00
3	12	9·94	2·06	0·43
4	10	9·69	0·31	0·01
5	6	8·74	−2·74	0·86
6	7	7·46	−0·46	0·03
7	6	6·11	−0·11	0·00
8	5	4·85	0·15	0·01
9	4	3·75	0·25	0·02
10	3	2·84	0·16	0·01
11	2	2·11	−0·11	0·01
12	2	1·55	0·45	0·13
13	1	1·12	−0·12	0·01
14 or more	3	2·57	0·43	0·07
Total	80	79·94		1·63

example 6 (see Table 3). These expected frequencies are compared with the observed frequencies in Table 7 and the χ^2 values for each frequency class are given in the fifth column of the table. As some expected frequencies were less than 1 (see Table 3), the values in one tail of the distribution were combined (Table 7). Therefore there are 15 frequency classes and the number of degrees of freedom $(v) = 15 - 3 = 12$.

The total χ^2 value of 1·63 is well below the 5% point of 21·03 ($Q = 0.050$ and $v = 12$) in Pearson & Hartley's (1966) Table 8. Therefore the model is a good fit to the original counts and agreement with a negative binomial distribution is accepted at the 95% probability level ($P > 0.05$).

5.2.2 Other tests for agreement with a negative binomial and tests for small samples (n < 50)

The χ^2 test for "goodness-of-fit" is also applied to small samples if the counts can be arranged in a frequency distribution. Other tests are available, and are used either to check the result of a χ^2 test or in situations where the χ^2 test cannot be used. The following tests are based on a comparison of observed and expected *moments* (first moment = arithmetic mean, second moment = variance, third moment: used with 2nd moment to calculate skewness):

Statistic U is the difference between the sample estimate of variance and the expected variance in a negative binomial.

$$U = s^2 - \left(\bar{x} + \frac{\bar{x}^2}{\hat{k}}\right)$$

where k is estimated from the frequency of zero counts (method 2 below).

Statistic T is the difference between the sample estimate of the third moment and the expected third moment.

$$T = \left(\frac{\Sigma x^3 - 3\bar{x}\,\Sigma x^2 + 2\bar{x}^2\,\Sigma x}{n}\right) - s^2\left(\frac{2s^2}{\bar{x}} - 1\right)$$

U and T have expected values of zero for perfect agreement with a negative binomial, but agreement is accepted if the value of U or T differs from zero by less than its standard error (defined in section 6.1). The standard error of U or T is easily calculated from Fig. 10 (exact formulae given by Anscombe 1950, Evans 1953), which also indicates the most efficient test for different values of $\bar{x}$ and $\hat{k}$. A large positive value of U or T indicates greater skewness than that described by the negative binomial and therefore the discrete log-normal distribution is a more suitable model; whereas a large negative value indicates less skewness and therefore models such as the Polya-Aeppli and Neyman distributions are more suitable (section 5.3).

5.2 NEGATIVE BINOMIAL DISTRIBUTION

The best estimate of k is always obtained by the maximum-likelihood method (section 3.2.3) when the counts of a small sample can be arranged in a frequency distribution. Alternative methods are often approximate and are used either as a first step towards the maximum-likelihood method, or when the more accurate method cannot be used.

(1) *Moment estimate of* k

$$\text{Estimate of } k = \hat{k} = \frac{\bar{x}^2}{s^2 - \bar{x}}$$

where $\bar{x}$ = arithmetic mean and s^2 = variance of sample. This formula is derived from the theoretical variance (see page 24) or from the statistics x' and y' where

$$x' = \bar{x}^2 - \frac{s^2}{n} \quad \text{and} \quad y' = s^2 - \bar{x}$$

The expectations of x' and y' are given exactly by:

$$x' = \mu^2 \quad \text{and} \quad y' = \frac{\mu^2}{k}$$

Therefore

$$\text{parameter } k = \frac{\mu^2}{y'} = \frac{x'}{y'}$$

The sample mean ($\bar{x}$) is a close estimate of the population mean (μ) in large samples, but not in small samples (see section 6.1). Therefore for large samples ($n > 50$),

$$\hat{k} = \frac{\mu^2}{y'} = \frac{\bar{x}^2}{y'} = \frac{\bar{x}^2}{s^2 - \bar{x}}$$

and for small samples,

$$\hat{k} = \frac{\mu^2}{y'} = \frac{x'}{y'} = \frac{\bar{x}^2 - s^2/n}{s^2 - \bar{x}}$$

where n = number of sampling units. As sample size (*i.e.* n) increases, s^2/n decreases and the above formula approaches

$$\left(\frac{\bar{x}^2}{s^2 - \bar{x}}\right)$$

The reciprocal of x'/y' forms the basis of a slope estimate of a common k for a series of samples (section 5.2.3).

(2) *Estimate of* k *from proportion of zeros*

Various values of $\hat{k}$ are tried in the following equation until the two sides are equal:

$$\hat{k} \log\left(1 + \frac{\bar{x}}{\hat{k}}\right) = \log\left(\frac{n}{f_0}\right)$$

where n = number of sampling units, f_0 = number of sampling units with no individuals = frequency of zero counts.

(3) *Transformation method for estimating* k

The purpose of transformations was explained in section 3.2.4 and suitable transformations for a negative binomial are given in Table 4. A rough estimate of k is first obtained by method 1 and then each count x is transformed to y where:

$$y = \log\left(x + \frac{k}{2}\right)$$

if rough estimate of $k = 2$ to 5, and $\bar{x} \geqslant 15$, or

$$y = \sinh^{-1}\sqrt{\frac{x + 0\cdot 375}{k - 0\cdot 75}}$$

if rough estimate of k is not less than 2, and $\bar{x} \geqslant 4$. The function $\sinh^{-1}$ is the reciprocal of the hyperbolic sine and is tabulated in Chambers' Mathematical Tables, Vol. 2, Table VIA (Comrie 1949). The expected variance of the transformed counts is independent of the mean and equal to (0·1886 trigamma k) for the first transformation, or (0·25 trigamma k) for the second transformation. Selected values of the function "trigamma k" are given in Table 8 (see also detailed Tables 13 to 16 of Davis 1935, Vol. 2). To estimate k, try different values of $\hat{k}$ in the appropriate transformation until the variance of the transformed counts equals the expected variance (see Table 8). These transformations and their expected variances are used to calculate confidence limits for small samples from a negative binomial (section 6.2.4).

Anscombe (1949, 1950) compared these three methods with the very efficient maximum-likelihood method. Method 1 is over 90% efficient for small values of $\bar{x}$ when $\hat{k}/\bar{x} > 6$, for large values of $\bar{x}$ when $\hat{k} > 13$, and for medium values of $\bar{x}$ when

$$\frac{(\hat{k} + \bar{x})(\hat{k} + 2)}{\bar{x}} \geqslant 15.$$

Therefore as k is often less than 4, method 1 is usually limited to samples with small means ($\bar{x} < 4$). Method 2 is over 90% efficient for

TABLE 8. EXPECTED VARIANCE OF TRANSFORMED COUNTS FROM A NEGATIVE BINOMIAL

0·1886 trigamma k for $\bar{x} \geqslant 15$

0·25 trigamma k for $\bar{x} \geqslant 4$

k		k		k		k	
2·0	0·1216	2·0	0·1612	6·5	0·0416	12·0	0·0217
2·1	0·1145	2·1	0·1517	6·6	0·0409	12·2	0·0214
2·2	0·1081	2·2	0·1432	6·7	0·0402	12·4	0·0210
2·3	0·1023	2·3	0·1356	6·8	0·0396	12·6	0·0207
2·4	0·0972	2·4	0·1288	6·9	0·0390	12·8	0·0203
2·5	0·0925	2·5	0·1226	7·0	0·0384	13·0	0·0200
2·6	0·0882	2·6	0·1170	7·1	0·0378	13·2	0·0197
2·7	0·0843	2·7	0·1118	7·2	0·0373	13·4	0·0194
2·8	0·0808	2·8	0·1071	7·3	0·0367	13·6	0·0191
2·9	0·0775	2·9	0·1028	7·4	0·0362	13·8	0·0188
3·0	0·0745	3·0	0·0987	7·5	0·0357	14·0	0·0185
3·1	0·0717	3·1	0·0950	7·6	0·0352	14·2	0·0183
3·2	0·0691	3·2	0·0916	7·7	0·0347	14·4	0·0180
3·3	0·0667	3·3	0·0884	7·8	0·0342	14·6	0·0177
3·4	0·0644	3·4	0·0854	7·9	0·0338	14·8	0·0175
3·5	0·0623	3·5	0·0826	8·0	0·0333	15·0	0·0172
3·6	0·0603	3·6	0·0800	8·1	0·0329	15·2	0·0170
3·7	0·0585	3·7	0·0775	8·2	0·0324	15·4	0·0168
3·8	0·0567	3·8	0·0752	8·3	0·0320	15·6	0·0166
3·9	0·0551	3·9	0·0730	8·4	0·0316	15·8	0·0163
4·0	0·0535	4·0	0·0710	8·5	0·0312	16·0	0·0161
4·1	0·0521	4·1	0·0690	8·6	0·0308	16·2	0·0159
4·2	0·0507	4·2	0·0672	8·7	0·0305	16·4	0·0157
4·3	0·0494	4·3	0·0654	8·8	0·0301	16·6	0·0155
4·4	0·0481	4·4	0·0638	8·9	0·0297	16·8	0·0153
4·5	0·0469	4·5	0·0622	9·0	0·0294	17·0	0·0152
4·6	0·0458	4·6	0·0607	9·1	0·0291	17·2	0·0150
4·7	0·0447	4·7	0·0593	9·2	0·0287	17·4	0·0148
4·8	0·0437	4·8	0·0579	9·3	0·0284	17·6	0·0146
4·9	0·0427	4·9	0·0566	9·4	0·0281	17·8	0·0145
5·0	0·0417	5·0	0·0553	9·5	0·0278	18·0	0·0143
		5·1	0·0542	9·6	0·0275	18·2	0·0141
		5·2	0·0530	9·7	0·0272	18·4	0·0140
		5·3	0·0519	9·8	0·0269	18·6	0·0138
		5·4	0·0509	9·9	0·0266	18·8	0·0137
		5·5	0·0498	10·0	0·0263	19·0	0·0135
		5·6	0·0489	10·2	0·0258	19·2	0·0134
		5·7	0·0479	10·4	0·0252	19·4	0·0132
		5·8	0·0470	10·6	0·0247	19·6	0·0131
		5·9	0·0462	10·8	0·0243	19·8	0·0130
		6·0	0·0453	11·0	0·0238	20·0	0·0128
		6·1	0·0445	11·2	0·0234		
		6·2	0·0438	11·4	0·0229		
		6·3	0·0430	11·6	0·0225		
		6·4	0·0423	11·8	0·0221		

$\bar{x} < 10$ when at least a third of the counts in a sample is zero (i.e. $f_0 > n/3$), but more zero counts are required for $\bar{x} > 10$. Therefore method 2 is chiefly limited to small means ($\bar{x} < 1$) or larger means with small values of $k(\hat{k} < 1)$. Method 3 is only appropriate when $\bar{x} > 4$, and then this method is more efficient than method 1. These efficiencies are for large samples and are probably lower for small samples.

The following key facilitates the choice of methods for small samples:

(A) *Counts can be arranged in a frequency distribution*—estimate k by method 1, then maximum-likelihood method (see example 6). Use χ^2 test for goodness-of-fit (see example 12) and make sure that all expected frequencies are at least 1 (section 4.1.3). These methods are often suitable for samples with n as low as 20, when $\bar{x}$ is fairly small (example 13B).

(B) *Counts cannot be arranged in a frequency distribution.* (B.1) Co-ordinates of $\bar{x}$ and $\bar{x}/k$ (k estimated by method 1) meet in U half of Fig. 10 and conditions for method 2 are fulfilled—estimate k by method 2 and calculate statistic U (example 13A).

(B.2) Co-ordinates of $\bar{x}$ and $\bar{x}/k$ meet in T half of Fig. 10 and method 2 cannot be used (not enough zero counts).

(B.2a) When $\bar{x} < 4$, accept estimate of k by method 1 and calculate statistic T (example 14).

(B.2b) When $\bar{x} > 4$, estimate k by method 3 and calculate statistic T (example 15). The slightly better estimate of k by method 3 will rarely justify the laborious calculations, especially when n is greater than about 15.

Example 13A. Method 2 for $\hat{k}$, and statistic U for testing agreement with a negative binomial distribution

A shovel sampler was used to take a small random sample of 20 sampling units from the bottom of a stony stream, and stonefly nymphs (*Protonemura meyeri*) were counted in each sampling unit (0.05m^2). The 20 counts are summarised in a frequency distribution thus:

x	0	1	2	3	7	8	9
f	7	3	4	2	1	2	1

where x is a particular count and f is the frequency of that particular count in the sample.

Number of sampling units $(n) = \Sigma f = 20$,
frequency of zero counts $(f_0) = 7$

5.2 NEGATIVE BINOMIAL DISTRIBUTION

Arithmetic mean $(\bar{x}) = \dfrac{\Sigma fx}{\Sigma f} = 2\cdot 4500,$

$\qquad$ variance $(s^2) = \dfrac{\Sigma (fx^2) - \bar{x}\Sigma fx}{n-1}$

$\qquad\qquad\qquad\quad = 9\cdot 20790$

Although the counts were arranged in a frequency distribution, their frequencies are rather irregular for $x > 3$. Therefore agreement with a negative binomial is tested by a comparison of observed and expected moments.

$\quad$ Rough estimate of k by method 1 $= \hat{k} = \dfrac{\bar{x}^2 - (s^2/n)}{s^2 - \bar{x}}$

$\qquad\qquad\qquad\qquad\qquad\qquad = \dfrac{2\cdot 45^2 - (9\cdot 20790/20)}{9\cdot 20790 - 2\cdot 45}$

$\qquad\qquad\qquad\qquad\qquad\qquad = 0\cdot 820$

Co-ordinates of $\bar{x}$ ($=2\cdot 45$) and $\bar{x}/k$ ($=2\cdot 45/0\cdot 8 = 3\cdot 1$) meet in U half of Fig. 10 and $f_0 > n/3$. Therefore k is estimated by method 2 thus: first try $\hat{k} = 0\cdot 8$ (estimate by method 1) in

$$\log\left(\dfrac{n}{f_0}\right) = k\log\left(1 + \dfrac{\bar{x}}{k}\right)$$

$$\log\left(\dfrac{n}{f_0}\right) = \log\left(\dfrac{20}{7}\right) = \log 2\cdot 85714 = 0\cdot 45591$$

$$k\log\left(1 + \dfrac{\bar{x}}{k}\right) = 0\cdot 8\log\left(1 + \dfrac{2\cdot 45}{0\cdot 8}\right) = 0\cdot 487080$$

As the latter value is higher than $\log(n/f_0)$, try $\hat{k} = 0\cdot 6$:—

$$k\log\left(1 + \dfrac{\bar{x}}{k}\right) = 0\cdot 6\log\left(1 + \dfrac{2\cdot 45}{0\cdot 6}\right) = 0\cdot 423672$$

This value is lower than $\log(n/f_0)$ and therefore $\hat{k}$ lies between $0\cdot 6$ and $0\cdot 8$. Therefore try $\hat{k} = 0\cdot 7$:—

$$k\log\left(1 + \dfrac{\bar{x}}{k}\right) = 0\cdot 7\log\left(1 + \dfrac{2\cdot 45}{0\cdot 7}\right) = 0\cdot 45725$$

This value is very close to $\log(n/f_0)$ and therefore $0\cdot 7$ is best estimate of k to one place of decimals. The iterative process can be continued for more places of decimals if a closer estimate of k is required. Agreement with a negative binomial distribution is now tested by the statistic U.

$$U = s^2 - \left(\bar{x} + \dfrac{\bar{x}^2}{k}\right) = 9\cdot 2079 - \left(2\cdot 45 + \dfrac{2\cdot 45^2}{0\cdot 7}\right) = -1\cdot 8171$$

$\quad$ Standard error of $U \simeq 2\left(\dfrac{10}{\sqrt{n}}\right) = \dfrac{20}{4\cdot 47} = 4\cdot 47$

from Fig. 10 for $\bar{x} = 2\cdot 45, \dfrac{\bar{x}}{k} = 3\cdot 5$, and $n = 20$.

Although this estimated standard error is not very reliable for n as low as 20, it does show that U is considerably less than its standard error and is even less than its standard error for a large sample ($n=100$). Therefore agreement with a negative binomial is accepted at the 95% probability level ($P>0.05$). It is impossible to state the exact probability level for the result, but it is at least 95% when U is less than its standard error.

Example 13B. χ^2 **test for goodness-of-fit**

The procedure for this test is fully described in examples 10, 11 and 12. Tables of Williamson & Bretherton (1963) were used to calculate the expected frequencies, and the nearest table was for "p"$=0.22$ (estimated from $\bar{x}$ and $\hat{k}$, see page 15), $\hat{k}=0.7$ and $\bar{x}=2.48$. Expected frequencies less than 1 were combined (see Table 9) and therefore the number of degrees of freedom $(v) = 6-3 = 3$. The total χ^2 value of 2·37 is well below the 5% point of 7·82 ($Q=0.050$ and $v=3$) in Pearson & Hartley's (1966) Table 8. Therefore the model is a good fit to the original counts and agreement with a negative binomial is accepted ($P>0.05$).

Although the U and χ^2 tests give similar results, the χ^2 test does not take full account of irregularities in the frequencies of the higher counts ($x > 3$). Therefore the χ^2 test is less sensitive than the second moment test, and may fail to detect departures from a negative binomial in small samples.

TABLE 9. χ^2 TEST FOR GOODNESS-OF-FIT OF A NEGATIVE BINOMIAL DISTRIBUTION

to a small sample. x, Obs., and Exp. are defined in Table 5.
Counts were obtained from example 13A.

x	Obs.		Exp.		χ^2
0	7		6·93		0·00
1	3		3·78		0·16
2	4		2·51		0·89
3	2		1·76		0·03
4	0		1·27		1·27
5	0		0·93		
6	0		0·69		
7	1	4	0·52	3·75	0·02
8	2		0·39		
9	1		0·29		
10 or more	0		0·93		
Total	20		20·00		2·37

Example 14. Method 1 for $\hat{k}$, and statistic T for testing agreement with a negative binomial distribution

The following counts of *B.rhodani* were obtained from a random sample of 20 sampling units:

x	0	1	2	3	4	5	6	7	8	9	10
f	1	3	3	3	3	1	2	1	1	1	1

5.2 NEGATIVE BINOMIAL DISTRIBUTION

Number of sampling units $(n) = 20$, arithmetic mean $(\bar{x}) = 4{\cdot}05$, variance $(s^2) = 8{\cdot}05$.

Rough estimate of k by method 1 $= \hat{k} = \dfrac{\bar{x}^2 - (s^2/n)}{s^2 - \bar{x}} = 4{\cdot}00$

Agreement with a negative binomial could be tested by χ^2 (goodness-of-fit) but the third moment test is probably more sensitive. Co-ordinates of $\bar{x}$ and $\bar{x}/\hat{k}$ meet in T half of Fig. 10 and there are too few zero counts for method 2. As $\bar{x} \simeq 4$, the estimate of k by method 1 is accepted and the statistic T is calculated thus:

$$T = \left(\frac{\Sigma x^3 - 3\bar{x}\,\Sigma x^2 + 2\bar{x}^2\,\Sigma x}{n}\right) - s^2\left(\frac{2s^2}{\bar{x}} - 1\right)$$

$$= \left[\frac{3441 - 12{\cdot}15(481) + 32{\cdot}805(81)}{20}\right] - 8{\cdot}05\left(\frac{16{\cdot}10}{4{\cdot}05} - 1\right)$$

$$= 12{\cdot}70275 - 23{\cdot}95125 = -11{\cdot}24850$$

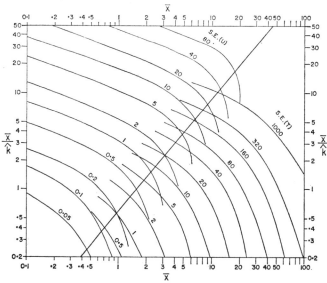

FIGURE 10. Standard errors of T and U for $n = 100$. For other values of n, multiply standard error by $\dfrac{10}{\sqrt{n}}$ (after Evans, 1953).

EXAMPLE: Co-ordinates of $\bar{x} = 2{\cdot}45$ and $\bar{x}/\hat{k} = 2{\cdot}45/0{\cdot}7 = 3{\cdot}5$ meet in U-half of figure, and near line for standard error of 2. This value must now be multiplied

by $\dfrac{10}{\sqrt{n}}$.

As $n = 20$, standard error of $U \simeq 2\left(\dfrac{10}{\sqrt{20}}\right) = 4{\cdot}47$ (see example 13A).

Standard error of $T \doteqdot 10 \left(\dfrac{10}{\sqrt{n}}\right) = \dfrac{100}{4\cdot 47} = 22\cdot 37$

from Fig. 10 for $\bar{x} = 4\cdot 05$, $\dfrac{\bar{x}}{\hat{k}} = 1\cdot 01$, and $n = 20$.

Therefore T is less than its standard error and agreement with a negative binomial is accepted at the 95% probability level ($P > 0\cdot 05$). As the estimated standard error is not very reliable for n as low as 20, a value of T close to the standard error would be suspect. Although the value of T does not indicate a significant departure from a negative binomial, the negative sign indicates a slight tendency towards a less skewed distribution *e.g.* Polya–Aeppli or Neyman (see section 5.3).

Example 15. Method 3 for $\hat{k}$, and statistic T for testing agreement with a negative binomial distribution

The following counts of *Gammarus pulex* were obtained from a small random sample of 10 sampling units ($n = 10$):—

$$4, \quad 5, \quad 8, \quad 14, \quad 14, \quad 15, \quad 15, \quad 19, \quad 28, \quad 36.$$

Arithmetic mean $(\bar{x}) = 15\cdot 8000$, variance $(s^2) = 99\cdot 06667$

Rough estimate of k by method 1 = $\hat{k} = \dfrac{\bar{x}^2 - (s^2/n)}{s^2 - \bar{x}} = 2\cdot 87910$

Co-ordinates of $\bar{x} = 15\cdot 8$ and $\bar{x}/\hat{k} \doteqdot 5\cdot 5$ meet in T half of Fig. 10, and therefore statistic T is calculated:

$$T = \left(\dfrac{\Sigma x^3 - 3\bar{x}\,\Sigma x^2 + 2\bar{x}^2\,\Sigma x}{n}\right) - s^2 \left(\dfrac{2s^2}{\bar{x}} - 1\right)$$

$$= \left[\dfrac{88406 - 47\cdot 4(3388) + 499\cdot 28(158)}{10}\right] - 99\cdot 06667 \left(\dfrac{198\cdot 13334}{15\cdot 8} - 1\right)$$

$$= 670\cdot 1040 - 1143\cdot 2373 = -473\cdot 1333$$

Standard error of $T \doteqdot 320 \left(\dfrac{10}{\sqrt{n}}\right) = \dfrac{3200}{3\cdot 1623} = 1011\cdot 936$

from Fig. 10 for $\bar{x} = 15\cdot 8$, $\dfrac{\bar{x}}{\hat{k}} \doteqdot 5\cdot 5$, and $n = 10$.

As T is about half its estimated standard error, agreement with a negative binomial is accepted at the 95% probability level ($P > 0\cdot 05$). Although the departure of T from its expected value of zero is not significant in this small sample, it is still quite large and the negative sign indicates a tendency towards a less skewed distribution.

As $\bar{x} > 4$, method 3 will give a slightly better estimate of k. The appropriate transformation is $y = \log(x + k/2)$; as estimate of k by method 1 is between 2 and 5, and $\bar{x} > 15$. First try $k = 2\cdot 9$ in transformation and therefore transformed counts are:

0·73640, 0·80956, 0·97543, 1·18894, 1·18894, 1·21618, 1·21618, 1·31069, 1·46909, 1·57345.

5.2 NEGATIVE BINOMIAL DISTRIBUTION

Mean of transformed counts = $\bar{y} = 1\cdot1685$, and variance of transformed counts = $0\cdot0703$. The expected variance for $k = 2\cdot9$ is $0\cdot0775$ (Table 8) and the actual variance of $0\cdot0703$ is between the expected variances for $k = 3\cdot1$ and $k = 3\cdot2$. Therefore try $k = 3\cdot1, 3\cdot2,$ and $3\cdot3$.

k	$\bar{y}$	Actual variance	Expected variance	Difference
3·1	1·1720	0·0692	0·0717	0·0025
3·2	1·1737	0·0686	0·0691	0·0005
3·3	1·1754	0·0681	0·0667	0·0014

As the least difference between actual and expected variance is for $k = 3\cdot2$, this value is the best estimate of k. If the estimate of k is required to more places of decimals, refer to tables of Davies (1935).

5.2.3. *Estimating a common* k *for a series of samples*

A common $k(k_c)$ can be calculated for a series of samples when there is no relationship between $\bar{x}$ (or log $\bar{x}$) and $1/k$ (Fig. 11A), but *not* when there is a definite trend or clustering (Fig. 11B). As k is an index of clumping in a Population (p. 23 and section 5.6), a stable k indicates that the level of clumping is a fairly constant characteristic of a species. A stable k is also essential in sequential sampling (section 8), or when the same transformation is required for a comparison of samples by t-test or analysis of variance (section 7.1.3).

A rough estimate of k_c is calculated from the statistics x' and y' which were used to derive a moment estimate of k (method 1) in section 5.2.2. These statistics are now calculated for each sample and plotted on a graph. It is assumed that a straight-line function expresses the average relation between the two variables, and this straight line is called the "regression line" of y' (ordinate) on x' (abscissa). The regression line passes through the origin ($y' = 0 = x'$) and has a slope $1/k_c$ which is given approximately by:

$$\frac{1}{k_c} = \frac{\Sigma y'}{\Sigma x'}$$

This rough estimate of k_c is usually adequate, but a more accurate estimate is obtained by the methods of Anscombe (1950), Bliss & Fisher (1953), or Bliss & Owen (1958).

A full explanation of a *regression line* is given in most standard statistical textbooks, *e.g.* Bailey (1959), Snedecor & Cochran (1967.) It must be noted that:

(1) a regression line is not a bad attempt to join up points on a graph;

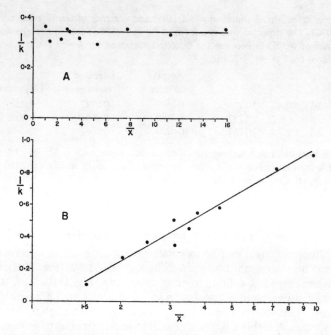

FIGURE 11. Relation of $1/k$ to $\bar{x}$ for (A) Ten samples from a population of *Gammarus pulex* (see example 16). Horizontal line is estimate of $1/k_c$. (B) Ten samples from a population of *Baetis rhodani*. Regression line shows that $1/k$ increases as $\bar{x}$ increases (note that $\bar{x}$ is on log scale).

TABLE 10. STATISTICS OF TEN RANDOM SAMPLES.

See example 16 for explanation of symbols. Sample 1 is from example 6 and sample 2 is from example 15.

Sample	$\bar{x}$	s^2	n	x'	y'	$\dfrac{1}{k}$
1	5·3125	13·5340	80	28·0535	8·2215	0·2931
2	15·80	99·06667	10	239·7333	83·2667	0·3473
3	7·70	27·55	10	56·5350	19·8500	0·3511
4	3·80	8·086	10	13·6314	4·2860	0·3144
5	2·80	5·326	10	7·3074	2·5260	0·3457
6	2·30	3·823	10	4·9077	1·5230	0·3103
7	1·05	1·421	20	1·0315	0·3710	0·3597
8	1·35	1·876	20	1·7287	0·5260	0·3043
9	3·05	6·124	20	8·9963	3·0740	0·3417
10	11·20	52·067	20	122·8366	40·8670	0·3327

5.2 NEGATIVE BINOMIAL DISTRIBUTION

(2) a regression line describes the average change in value of one variable (*dependent* variable y) for a unit change in another related variable (*independent* variable x, also called *regressor* variable);

(3) a regression line is always fitted in such a way that the sum of squares of deviations of the dependent variable from the line is minimised;

(4) one or both of the variables (x and y) may be transformed to logarithms, so that a straight regression line is obtained (see Fig. 11B).

Example 16. Calculation of a common k

Counts of *Gammarus pulex* were obtained from 10 random samples over a period of several months. The mean ($\bar{x}$), variance (s^2), and number of sampling units (n) for each sample are given in Table 10, together with the statistics x', y', and $1/k$.

$$x' = \bar{x}^2 - \frac{s^2}{n},$$
$$y' = s^2 - \bar{x},$$
$$\frac{1}{k} = \frac{y'}{x'} = \frac{s^2 - \bar{x}}{\bar{x}^2 - s^2/n}$$

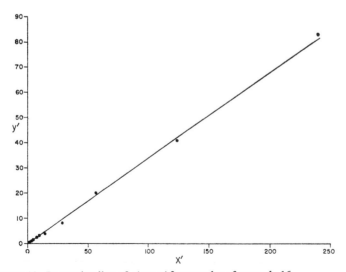

FIGURE 12. Regression line of y' on x' for samples of example 16.

There is no relationship between $\bar{x}$ and $1/k$ (see Fig. 11A), and therefore the calculation of a common k (k_c) is justified. As the values of x' and y' are used to estimate k_c, enough places should be recorded to avoid subsequent rounding errors. The estimate of the slope of the regression line is:

$$\frac{1}{k_c} = \frac{\Sigma y'}{\Sigma x'} = \frac{164 \cdot 5112}{484 \cdot 7614} = 0 \cdot 3394$$

The regression line of y' on x' is fitted by joining the lowest and highest points of the line (see Fig. 12). Lowest point = 0. Highest point = expected value of y' for maximum value of x'

$$= \frac{\text{maximum } x'}{k_c} = 239 \cdot 7333(0 \cdot 3394) = 81 \cdot 3655$$

As most of the points lie close to the regression line, the latter appears to be a good fit to the data. Therefore $0 \cdot 3394$ is accepted as a good estimate of $1/k_c$ and the estimate of a common $k = 1/0 \cdot 3394 = 2 \cdot 9464 = 2 \cdot 95$. The stability of k implies that the degree of contagiousness in the Population was relatively constant over several months, despite the variation in sample means.

5.3 OTHER CONTAGIOUS FREQUENCY DISTRIBUTIONS

The negative binomial has been successfully fitted to many contagious invertebrate Populations and appears to have a wide application. Few of these studies deal with freshwater invertebrates, but the negative binomial was fitted to samples of zooplankton (Comita & Comita 1957, Colebrook 1960) and to samples of benthic invertebrates from streams in the Lake District. The negative binomial will probably fit samples from other freshwater habitats, but this assumption cannot be tested until there is more information on the dispersion of freshwater invertebrates. There is always a possibility that another model fits the sample better than the negative binomial, and therefore several contagious frequency distributions are briefly described below.

Skewness refers to the bilateral asymmetry of a frequency distribution, and is positive when the distribution tails off among the higher counts (*e.g.* Fig. 3) or negative when it tails off among the lower counts. Anscombe (1950) ranked the following contagious distributions in order of increasing skewness: Thomas (1949), Neyman (1939) Type A, Polya-Aeppli (Polya 1931), negative binomial, and discrete log-normal. All five distributions are positively skewed (Fig. 13) with a variance greater than the mean. Three distributions are less skewed than the negative binomial and are based on hypotheses similar to model 3 for the negative binomial (p. 51). Clumps of individuals are distributed at random, and the

5.3 OTHER CONTAGIOUS FREQUENCY DISTRIBUTIONS

numbers of individuals in the clumps are distributed independently in either a Poisson distribution (Thomas, Neyman), a geometric distribution (Polya-Aeppli), or a logarithmic distribution (negative binomial).

The Thomas and Neyman distributions are very similar and may both fit the same sample. Neyman's distribution is intended to describe the dispersion of insect larvae which have recently hatched from randomly distributed clumps of eggs. There is an arbitrary limit to clump size which depends upon the distance crawled by the larvae. The distribution is defined by two parameters, one equal to the mean number of clumps per sampling unit (m_1) and one equal to the mean number of larvae per clump (m_2).

$$m_1 = \frac{\mu}{m_2} \quad \text{and} \quad m_2 = \frac{\sigma^2}{\mu} - 1$$

where m_1 and m_2 are the parameters of two independent Poisson series. In the double Poisson distribution of Thomas, one must be subtracted from the numbers in each clump before they follow a Poisson series with mean $= m_2 - 1$. The most useful property of these

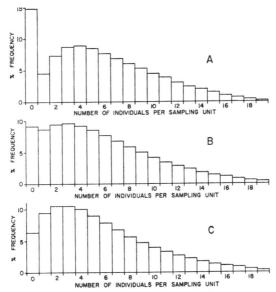

FIGURE 13. Three frequency distributions for the same mean (=6) and variance (=24). (A) Neyman Type A with modes at $x = 0$ and $x = 4$. (B) Polya-Aeppli with modes at $x = 0$ and $x = 3$. (C) Negative binomial with single mode at $x = 2$ to 3. Note that skewness increases from A to C.

two models is their capacity to describe both polymodal and unimodal frequency distributions, whereas the negative binomial is always unimodal (*mode* = peak in frequency distribution, *cf.* Figs. 13A and C). Examples of the application of these models are given by Beall (1940), Thomson (1952), and Hartenstein (1961).

The Polya-Aeppli distribution arises from the simultaneous, random colonisation of a habitat by parent individuals. The parents produce clusters of offspring and the numbers of individuals per cluster have a geometric distribution. This model closely parallels the negative binomial, but may have two modes under certain conditions (Fig. 13B). As it arises from a rather specialised set of circumstances, the Polya-Aeppli distribution cannot have the general application of the negative binomial. Agreement with the Neyman and Polya-Aeppli distributions is tested by χ^2 or by a comparison of observed and expected moments (for methods see Evans 1953).

The discrete log-normal distribution is more skewed than the negative binomial, but also has only one mode. When the logarithms of counts follow a normal frequency distribution, the original counts of the sample must follow a discrete log-normal distribution. The log-normal distribution is often a useful approximate model, and is the basic assumption for a straight logarithmic transformation (section 3.2.4). Cassie (1962) has derived a Poisson-log-normal distribution which has a skewness intermediate between that of the negative binomial and the discrete log-normal. The model for the Poisson-log-normal is similar to model 4 for the negative binomial (page 52), but the parent distribution is log-normal instead of a Pearson Type III distribution.

Cole (1946) suggested that contagious distributions should be considered as a superimposition of independent random distributions of groups of $1, 2, 3, 4, \ldots n'$ individuals. The parameters of the distribution represent the density of single individuals, pairs, singles and pairs, etc. As six parameters are often used, it is not surprising that this compound distribution is a better fit than other two-parameter distributions. The method is laborious and will only be justified when a detailed analysis of spatial distribution is required.

5.4 Effect of quadrat size

The effect of quadrat size (*i.e.* the size of the sampling unit) on the detection of non-randomness was briefly discussed in section 4.1. If quadrat size is steadily increased, the apparent dispersion of a

5.4 EFFECT OF QUADRAT SIZE

contagious Population may be random, contagious, and finally regular. This is illustrated by a contagious Population with regularly distributed clumps (Fig. 14). Four quadrat sizes (A, B, C, D) are used to sample this Population. The dispersion is apparently random or slightly contagious with quadrat A $(s^2 \simeq \bar{x})$, definitely contagious with quadrat B (each quadrat contains very few or very many individuals, and therefore $s^2 > \bar{x}$), random with quadrat C $(s^2 \simeq \bar{x})$, and finally regular with quadrat D (clumping will only affect the dispersion within the quadrat and not the number it contains; therefore, each quadrat contains approximately the same number of individuals and $s^2 < \bar{x}$). This final decline in variance with increasing quadrat size (*e.g.* quadrats C and D) only occurs when the clumps are regularly distributed (as in Fig. 14), and is less pronounced when the distribution of the clumps is random or contagious.

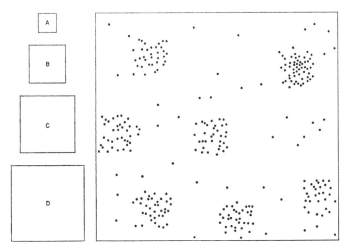

FIGURE 14. Four quadrat sizes (A, B, C, D) and a contagious distribution with regularly distributed clumps.

When the dispersion of a Population is truly random, the variance will steadily increase with quadrat size so that the variance to mean ratio is always unity $(s^2 \simeq \bar{x})$. This result for several quadrat sizes is strong evidence for a random dispersion, but alternative explanations are: (1) the largest quadrat size is smaller than the mean size of the clumps, *i.e.* the dispersion is contagious with very large clumps of individuals; or (2) the smallest quadrat size is larger than the mean size of the clumps which are regularly distributed.

The first explanation is unlikely unless the density of the Population is very high, whereas the second explanation is often applicable to low density Populations when the individuals in the Population are arranged in small compact groups. It is usually easy to check the first explanation by sampling with a larger quadrat. The second explanation is often impossible to check as the size of the smallest quadrat is usually limited by physical factors, *e.g.* stone size often limits quadrat size to about 0·05 m² for small streams in the Lake District. Therefore, it is preferable to use two or more quadrat sizes, and one bottom sampler should have the smallest possible quadrat size, *e.g.* a Macan shovel sampler removes 0·05 m² of bottom and is, therefore, suitable for stony streams.

The relationship between quadrat size and clump size can be used to detect the actual scale of clumping in a Population. A series of samples is taken and a different quadrat size is used for each sample. The variance of each sample is plotted against the appropriate quadrat size, and maximum variance occurs when quadrat size and clump size are approximately equal. This technique is used to analyse spatial patterns in plant ecology (see references in Greig-Smith 1964, pp. 85 to 93). Morisita (1959) has developed a similar technique for the analysis of pattern and uses his *index of dispersion* I_δ (see section 5.6.4). This index is calculated for a series of quadrats where each quadrat size is double the size of the previous quadrat. If the smallest quadrat is q cm², the ratio

$$\frac{I_\delta \text{ for quadrat } q}{I_\delta \text{ for quadrat } 2q}$$

is plotted against quadrat size $2q$. This process is then repeated for quadrats $2q$ and $4q$, $4q$ and $8q$, $8q$ and $16q$, etc. Peaks occur in the graph when quadrat size and clump size are approximately equal. In one example (Fig. 15), a definite peak occurs at quadrat size $8q$

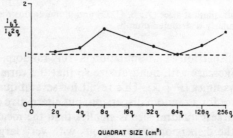

FIGURE 15. Changes in $\dfrac{I_\delta \text{ for } q}{I_\delta \text{ for } 2q}$ with increasing quadrat size.

and another peak may exist at a quadrat size larger than $256q$. Therefore, the Population consists of small clumps with a mean size of about $8q$ cm^2, and a large aggregation of these clumps covers an area greater than $256q$ cm^2. Larger quadrats are needed to determine the size of this large aggregation.

5.5 Taylor's Power Law

The power law (Taylor 1961) states that: the variance (σ^2) of a population is proportional to a fractional power of the arithmetic mean (μ):

$$\sigma^2 = a\mu^b \quad \text{and therefore} \quad \log \sigma^2 = \log a + b \log \mu$$

where a and b are population parameters.

Parameter a depends chiefly upon the size of the sampling unit. Parameter b is an index of dispersion and varies continuously from 0 for a regular distribution to infinity for a highly contagious distribution ($a = b = 1$ when dispersion is random). Once b is estimated, a common transformation can be applied to the original counts and then methods associated with the normal distribution are applicable (see section 3.2.4). The appropriate transformation is to replace each count by x^p, where $p = 1 - b/2$. Some of these transformations are the same as those derived by other methods; e.g. for Poisson, $b = 1$ and $p = 0.5$, therefore use square roots ($x^{0.5} = \sqrt{x}$); for log-normal, $b = 2$ and $p = 0$, therefore use logarithms. Healy & Taylor (1962) gives tables of transformations for $p = 0.2$, 0.4, 0.6, 0.8, and for the negative powers. Therefore, the applications of parameter b are similar to those of a common k (section 5.2.3). The advantages of b over k_c are: (1) the power law covers a wider range of distributions than the negative binomial; and (2) the transformations derived from b are often easier to apply than those derived from the negative binomial.

The variances (s^2) of a series of samples are plotted against their corresponding means ($\bar{x}$) on a log/log scale. Either the logarithms of s^2 and $\bar{x}$ are plotted on ordinary graph paper, or the original arithmetic values are plotted on double-log graph paper. This paper has the log scale on both axes and is available with one or more *cycles* (each cycle is a multiple of 10, e.g. 0.1 to 1.0, 1 to 10, 10 to 100). A log scale was used in Figs. 6, 10, 11B. The use of different graph papers is explained in detail by Lewis & Taylor (1967, pp. 36–78). Parameters a and b are estimated by the intercept (a) and regression coefficient (b) of the regression line of log s^2 on log $\bar{x}$. This regression line can be fitted graphically, or the estimates of a and b can be calculated.

Example 17. Application of Taylor's power law
(a) Graphical method

The means and variances were obtained from the 10 samples of example 16 (see Table 10). Variance (s^2) was plotted against mean ($\bar{x}$) on a double-log scale (Fig. 16). A straight line was fitted to the points by eye and this line is the regression line of $\log s^2$ on $\log \bar{x}$ ($\log \bar{x}$ is always on the abscissa). The value of a is always read off the s^2 axis for $\bar{x} = 1$, and the regression coefficient b is given by the tangent of the angle of elevation (β) for the regression line. Therefore, estimates of both parameters are (from Fig. 16):

$$a = 1 \cdot 1 \text{ and}$$
$$b = \tan \beta = \tan 57°30' = 1 \cdot 57$$

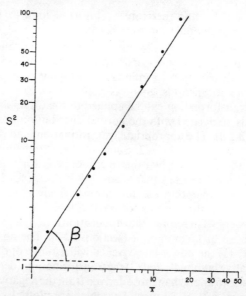

FIGURE 16. Regression line of $\log s^2$ on $\log \bar{x}$. The regression line was fitted by eye. Value of s^2 for $\bar{x} = 1$ is $a = 1 \cdot 1$. Angle of elevation $= \beta = 57°30'$.

Therefore the power law for these samples is:

$$\log \sigma^2 = \log a + b \log \mu = \log 1 \cdot 1 + 1 \cdot 57 \log \mu$$
$$\sigma^2 = a\mu^b = 1 \cdot 1 \mu^{1 \cdot 57}$$

As $\qquad b = 1 \cdot 57$,

$$p = 1 - \frac{b}{2} = 1 - 0 \cdot 785 = 0 \cdot 215$$

and therefore the appropriate transformation is $x^p = x^{0 \cdot 2}$. The transformed counts can be read off from Table 1 of Healy & Taylor (1962), and then "normal" methods can be applied, *e.g.* t-tests and analysis of variance.

In this example, the line was easily fitted by eye, but the following method is used when there are more points:
(1) Using vertical lines, divide the points into three roughly-equal groups.
(2) Find the medial points of the two end groups by drawing a medial cross. This cross divides the points in each group equally between upper and lower segments, and left and right segments.
(3) The regression line is estimated by joining these two medial points.
(4) This regression line can be slightly improved by drawing a second line which is parallel to the first and passes through the mean of $\log s^2$ and $\log \bar{x}$. This new line will change a but not b.

(b) Calculation of a and b

More accurate estimates of the parameters a and b are given by:

$$b = \frac{\Sigma(x-\bar{x})(y-\bar{y})}{\Sigma(x-\bar{x})^2} = \frac{n\Sigma(xy) - (\Sigma x)(\Sigma y)}{n\Sigma(x^2) - (\Sigma x)^2}$$

$$\log a = \bar{y} - b\bar{x} = \frac{\Sigma y - b\Sigma x}{n}$$

where $x = \log \bar{x}$ and $y = \log s^2$ for each sample, and n = number of samples. A full explanation of these methods is given in most standard textbooks, e.g. Bailey (1959).

In this example, $b = 1.5748$,
$\log a = 0.0447$ and therefore $a = 1.1084$

Therefore a more accurate statement of the power law is:

$$\sigma^2 = 1.1084\mu^{1.5748}$$

The rough estimates obtained by the graphical method are very close to these accurate estimates of a and b. Therefore the laborious calculations of the second method are rarely justified.

5.6 INDICES OF DISPERSION

Many different indices have been proposed to compare the different patterns of dispersion in Populations. The ideal index of dispersion should possess the following attributes:
(1) It should provide real and continuous values over the range from maximum regularity (equal numbers in each sampling unit), through randomness ($s^2 = \bar{x}$), to maximum contagion (all individuals are in one sampling unit).
(2) It should not be influenced by variation in the size of the sampling unit (quadrat size), the number of sampling units (n), the sample mean ($\bar{x}$), and the total numbers in the sample (Σx).

(3) It should be easy to calculate from large amounts of data.
(4) It should enable differences between samples to be tested for significance.

There is no perfect index of dispersion which fulfils all these conditions. The following indices are most frequently used.

5.6.1 Indices based on the variance to mean ratio

The application and calculation of the variance to mean ratio were fully explained in section 4.1.2. There are many variants of this ratio, and several variants are compared in Table 11. The values in the last column (maximum contagion) indicate that nearly all these indices are strongly influenced by the number of individuals in the sample ($\Sigma x = n\bar{x}$). Therefore the variance to mean ratio can only be used as a comparative index of dispersion when Σx, $\bar{x}$ and n have the same values in each sample. This is rarely possible. Therefore the variance to mean ratio is a good statistical test for agreement with a Poisson series (section 4.1.2), but is not a good measure of the degree of clumping in a Population. This criticism is also applicable to most derivatives of the variance to mean ratio. The Charlier coefficient (Table 11) is independent of the sample mean and total numbers when Σx is large, but is always affected by the number of sampling units (n). Green's coefficient (Table 11) is the only variant which is independent of variation in n, $\bar{x}$ and Σx. This index is therefore

TABLE 11. VALUES OF VARIOUS INDICES OF DISPERSION
at maximum regularity, randomness, and maximum contagion. n = number of sampling units, $\bar{x}$ = sample mean, s^2 = sample variance, $n\bar{x} = \Sigma x$ = total numbers in the sample.

Index	Max. regularity	Random	Max. contagion
$\dfrac{s^2}{\bar{x}}$ (Section 4.1.2)	0	1	Σx
$\dfrac{s^2(n-1)}{\bar{x}}$ (Section 4.1.2)	0	$n-1$	$\Sigma x(n-1)$
$\dfrac{s^2}{\bar{x}} - 1$ (David & Moore 1954)	-1	0	$\Sigma x - 1$
$\dfrac{s}{\sqrt{\bar{x}}}$ (Index of Lexis)	0	1	$\sqrt{\Sigma x}$
$100\dfrac{\sqrt{s^2 - \bar{x}}}{\bar{x}}$ (Charlier coefficient)	imaginary	0	$100n\left(1 - \dfrac{1}{\Sigma x}\right)$
$\dfrac{(s^2/\bar{x}) - 1}{\Sigma x - 1}$ (Green 1966)	$-\dfrac{1}{\Sigma x - 1}$	0	1

5.6 INDICES OF DISPERSION

suitable for comparisons of contagious distributions, and ranges from 0 for random dispersion to 1 for maximum contagion. Green (1966) pessimistically concludes that any coefficient of non-randomness should not be calculated when $n < 50$, and "where high positive contagion is present, at least several hundred samples should be used"!!

The variance to mean ratio was criticised by Skellam (1952) on the grounds that it is dependent on the size of the sampling unit (quadrat size). As the detection of non-randomness by quadrat sampling is always dependent on the size of the sampling unit (see section 5.4), this criticism is also applicable to other indices of dispersion.

5.6.2 k in the negative binomial

This index is frequently used as an index of dispersion, and is compared with other indices in Table 12. The calculation of k is fully described in sections 3.2.3 and 5.2.2. Use of k requires agreement with the negative binomial. The disadvantages of this index are, (1) it is not independent of the number of sampling units (n), (2) it goes to $\pm$ infinity at randomness (Table 12), (3) the value of k is often influenced by the size of the sampling unit. Therefore comparisons of the level of clumping can only be made with k when n and quadrat size are the same in each sample. As k is inversely proportional to the level of clumping, its reciprocal $1/k$ is often a more convenient index of dispersion but is also affected by the number of sampling units (Table 12).

TABLE 12. VALUES OF VARIOUS INDICES OF DISPERSION

Index	Max. regularity	Random	Max. contagion
k (negative binomial)	$-\bar{x}$	$\pm \infty$	$\dfrac{1}{n}\left(1 + \dfrac{1}{\Sigma x - 1}\right)$ (tends to zero)
$\dfrac{1}{k}$	$-\dfrac{1}{\bar{x}}$	0	$n - \dfrac{1}{\bar{x}}$
b in power law	0	1	∞
I_δ (Morisita)	$1 - \dfrac{n-1}{\Sigma x - 1}$	1	n
$\dfrac{\Sigma(x^2)}{(\Sigma x)^2}$ (Cole 1946b)	$\dfrac{1}{n}$	$\dfrac{1}{n} + \dfrac{n-1}{n}\left(\dfrac{1}{\Sigma x}\right)$	1
$\dfrac{1}{45}\tan^{-1}\left(\dfrac{s^2}{\sigma^2}\right) - 1$ (Lefkovitch 1966)	-1	0 (when Poisson, $\sigma^2 = \bar{x}$)	$+1$

5.6.3 b *of Taylor's power law*

The parameter b (section 5.5) is a measure of the degree of clumping in a Population and is often fairly constant for a species. As it has a wide range and is apparently independent of n, $\bar{x}$ and Σx (Table 12), this index is of value in interspecific comparisons (see examples in Fig. 17). Calculation of b requires several estimates of s^2 and $\bar{x}$, and therefore b cannot be calculated for one sample. As quadrat size affects the relationship between these two statistics (see section 5.4), it will also affect the value of b.

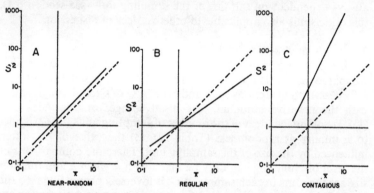

FIGURE 17. Regression lines of log s^2 on log $\bar{x}$ for different patterns of dispersion. (A) Near-random distribution, $\sigma^2 = 1 \cdot 5\mu^{1 \cdot 0}$; (B) Regular distribution, $\sigma^2 = 1\mu^{0 \cdot 7}$; (C) Contagious distribution, $\sigma^2 = 10\mu^{2 \cdot 0}$. The individual points around the regression line are omitted.

5.6.4 *Morisita's index of dispersion*

Morisita (1959) has developed the following index of dispersion:

$$I_\delta = n \frac{\Sigma[x(x-1)]}{\Sigma x(\Sigma x - 1)} = n \frac{\Sigma(x^2) - \Sigma x}{(\Sigma x)^2 - \Sigma x}$$

This index is independent of the sample mean ($\bar{x}$) and total numbers in the sample (Σx), but is a strong function of the number of sampling units (n) at both ends of its range (Table 12). Therefore I_δ is a good comparative index of dispersion when each sample contains the same number of sampling units. The index equals one for a random distribution, is greater than one for a contagious distribution, and is less than one for a regular distribution. Departures from randomness are judged significant ($P < 0.05$) when:

$$I_\delta(\Sigma x - 1) + n - \Sigma x$$

is outside the appropriate 5% significance levels of χ^2 for $n-1$ degrees of freedom (see Table 8 in Pearson & Hartley 1966).

Morisita (1959) has investigated changes in I_δ with different sizes of sampling unit (quadrat size), and these changes are summarised in Fig. 18. I_δ is strongly influenced by quadrat size when the dispersion of a Population is regular (Fig. 18A). When the dispersion is contagious and individuals are randomly distributed in each clump, I_δ is fairly stable until clump size and quadrat size are approximately equal (Figs. 18B, C). This decrease in I_δ can be used to estimate the mean size of the clumps (see page 70). The index is less stable when the intra-clump distribution is uniform (Fig. 18D).

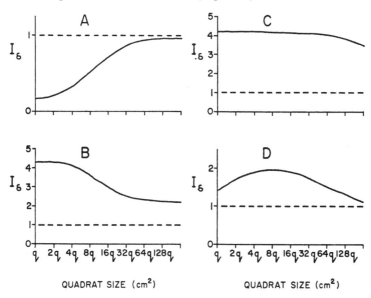

FIGURE 18. Changes in I_δ with increasing quadrat size. (A) Regular distribution. (B) Contagious distribution with small clumps. (C) Contagious distribution with large clumps. (D) Contagious distribution with uniform intra-clump distribution. The broken horizontal lines indicate a random distribution ($I_\delta = 1$). Smallest quadrat size = q sq. cm.

5.6.5 *Other indices of dispersion*

Cole's index has only an empirical basis and is not independent of n (Table 12). Lefkovitch (1966) has developed an index of dispersion from the observed variance (s^2) and the theoretical variance ($\sigma^2 = \mu$ for a Poisson series). s^2 is plotted against σ^2 on the same scale and a straight line drawn to the origin. The angle between this line and the σ^2 axis is $\tan^{-1} s^2/\sigma^2$ and this equals 45° when $s^2 = \sigma^2$; *e.g.* when

$\sigma^2 = \bar{x}$ (Poisson), angle = 45° for random, < 45° for regular, > 45° (to 90°) for contagious. The formula in Table 12 changes these angles to a new scale ranging from -1 to $+1$. Lefkovitch provides a table for testing the significance of departures from randomness.

Lloyd (1967) has developed an index of *mean crowding* ($\overset{\cdot}{m}$), which defines the mean number per individual of other individuals in the same quadrat:

$$\overset{\cdot}{m} = \mu + \left(\frac{\sigma^2}{\mu} - 1\right)$$

The ratio of mean crowding to mean density is a suitable measure of patchiness. When the data agree with the negative binomial, this measure of patchiness is:

$$\frac{\overset{\cdot}{m}}{\mu} = 1 + \frac{1}{k}$$

and therefore the sample estimate of mean crowding is:

$$\overset{\cdot}{x} = \bar{x}\left(1 + \frac{1}{\hat{k}}\right)$$

This index has been developed further by Iwao (1968), and Iwao & Kuno (1968).

All the above indices are affected to a greater or lesser extent by quadrat size, and it is often impossible to detect non-randomness when clumps of individuals are very small (section 5.4). These problems can be overcome by using indices which are based on nearest neighbour measurements (Greig-Smith 1964, Southwood 1966). As these techniques require stationary, easily seen animals, they are limited to sedentary invertebrates or mobile animals which can be photographed.

5.7 Summary

The spatial dispersion of a Population is fundamental to a study of its ecology. Different species have different environmental requirements and behaviour patterns. Therefore the dispersion of each species follows a characteristic pattern. The dispersion of a Population is seldom, if ever, random but may be close to random when the density of the Population is low, or when clumping is not detected by a large quadrat size (sections 4.1, 5.4).

Therefore the dispersion is usually contagious, and the level of clumping in the Population will affect the final spatial pattern (Figs. 7C, 9). If a mathematical model fits samples from a population, then:

(1) the dispersion of the Population can be described in mathematical terms;

5.7 SUMMARY

(2) errors of population parameters can be estimated (section 6);
(3) temporal and spatial changes in density can be compared (section 7);
(4) the effect of environmental factors can be assessed.

There are many mathematical models for contagious distributions (see section 5.3), and the negative binomial distribution appears to be the most flexible (section 5.2). Agreement with the negative binomial is tested by χ^2 (for large samples), or by comparing moments (see key on page 58).

Another useful model is Taylor's power law which covers a wider range of distributions than the negative binomial (section 5.5). There are many indices of dispersion (section 5.6), and the most useful are:

(1) k (or $1/k$), when the negative binomial is a suitable model;
(2) b of Taylor's power law for a series of samples;
(3) Morisita's index, when a detailed analysis of the pattern of dispersion is required.

VI THE PRECISION OF A SAMPLE MEAN

6.1 STANDARD ERROR OF THE MEAN

In chapter 2, we noted that the arithmetic mean of a population (parameter μ) is estimated by the arithmetic mean of a sample from the population (statistic $\bar{x}$). As the estimated population mean is often used to determine the total numbers in the sampling area, it is important to know the accuracy of this estimate. For example, the mean number of nymphs in 100 cm² is 10·125 (see example 3) and therefore the estimated population mean is 10·125 nymphs per 100 cm². If the sampling units were taken from a section of stream with a total area of 30 m² ($= 300,000$ cm²), then the estimate of the total number of nymphs in the sampling area

$$= \left(\frac{300,000}{100}\right) 10 \cdot 125 = 30,375$$

It would be remarkable if this estimate agreed with the actual number of nymphs in the sampling area, and therefore it is important to know the *error* of the estimated population mean.

The true population mean (μ) has only one value at the time of sampling, whereas the value of the sample mean ($\bar{x}$) varies from one sample to the next. For example, six samples may have means of 9·51, 10·74, 9·82, 10·20, 10·52, 10·13, and all these values can be used as estimates of the unknown population mean. If it was possible to take a large number of samples from the same population and arrange the sample means in a frequency distribution, the latter would tend to normality even if the counts of each sample followed non-normal distributions, *e.g.* positive binomial, Poisson, and negative binomial distributions. This important result is summarised by the *central-limit theorem* which states that: the means of large random samples from the same population are approximately normally distributed with a mean equal to the true population mean (μ) and a variance which is related to the population variance (σ^2) by the simple formula:

$$\text{variance of sample means} = \frac{\sigma^2}{n} \quad \text{and hence}$$

$$\text{standard deviation of sample means} = \sqrt{\frac{\sigma^2}{n}}$$

Therefore if the number of sampling units (n) is sufficiently large

6.2 CONFIDENCE LIMITS OF THE MEAN

($n \geq 30$), the sampling distribution of the sample means: (a) is approximately normal, (b) has mean μ, and (c) has standard deviation $\sqrt{\sigma^2/n}$ which is estimated by $\sqrt{s^2/n}$. The standard deviation of the sample means is usually called the *standard error* of the mean and indicates the amount of error in the sample mean ($\bar{x}$) when it is used to estimate the population mean (μ). Therefore the estimate of μ is often written as $\bar{x} \pm$ standard error ($\bar{x} \pm \sqrt{s^2/n}$). As the size of the sample (*i.e.* n) increases, the standard error decreases, and the estimate of the population mean approaches μ. For the counts in example 3, $\bar{x} = 10 \cdot 125$ and the standard error of the mean

$$= \sqrt{\frac{s^2}{n}} = \sqrt{\frac{7 \cdot 465}{80}} = 0 \cdot 306$$

whereas for the first 8 counts only, $\bar{x} = 10 \cdot 500$ and standard error of the mean

$$= \sqrt{\frac{12 \cdot 857}{8}} = 1 \cdot 268$$

which is about four times the standard error of the 80 counts.

6.2 CONFIDENCE LIMITS OF THE MEAN

Confidence limits are often used instead of the standard error and are easier to interpret. These limits define the upper and lower values of a range within which the true population mean lies, e.g. the 95% confidence limits indicate that the odds are 95 to 5 (or 19 to 1) that the population mean lies between these limits.

6.2.1 *Normal approximation with large samples*

If a large number of sampling units form the sample ($n \geq 30$) and the sampling units are taken at random, then the central-limit theorem is applicable. Therefore the sample mean is one of many possible sample means which are normally distributed around the true population mean. In a normal distribution, 95% of the values lie within a range of 1·96 standard deviations on each side of the mean (see standard textbooks). Therefore 95% of the possible sample means will lie within a range of 1·96 standard errors on each side of the population mean, and the odds are 19 to 1 ($P = 0·95$) that one sample mean will lie within this range. There are similar odds ($P = 0·95$) that the population mean (μ) will lie within 1·96 standard errors on each side of the sample mean ($\bar{x}$), and the limits of this range are the 95% confidence limits of the mean. The lower confidence limit is

THE PRECISION OF A SAMPLE MEAN

$\bar{x} - 1.96$ standard errors, and the upper limit is $\bar{x} + 1.96$ standard errors; or more exactly the 95% confidence limits are given thus:

$$\bar{x} - t\sqrt{\frac{s^2}{n}} \quad \text{to} \quad \bar{x} + t\sqrt{\frac{s^2}{n}}$$

where $\sqrt{s^2/n}$ is the standard error and t is found in *Student's t-distribution*. Student's t must always be used when the population variance (σ^2) is not known and has to be estimated by the sample variance (s^2).

The value of t depends upon the number of degrees of freedom $(n-1)$, and increases steadily as the number of degrees of freedom declines (Table 13). As the values of t for the 95% limits are all close to 2, the latter value is a close approximation for most purposes (with $n > 30$, see Table 13).

TABLE 13. VALUES OF t FOR 95% AND 99% CONFIDENCE LIMITS
with different degrees of freedom $(n-1)$.

95% limits		99% limits	
$n-1$	t	$n-1$	t
30	2.04	30	2.75
35	2.03	35	2.72
40	2.02	40	2.70
45 to 54	2.01	45	2.69
55 to 70	2.00	50	2.68
71 to 98	1.99	55	2.67
99 to 165	1.98	58 to 64	2.66
>165	1.97	65 to 73	2.65
∞	1.96	74 to 85	2.64
		86 to 103	2.63
		104 to 129	2.62
		130 to 174	2.61
		>175	2.60
		∞	2.58

Although 95% confidence limits are suitable for most estimates of the population mean, there are also values of t for 99% confidence limits (see Table 13), and any other level of probability that may be required. With 99% limits, the odds are 99 to 1 that the population mean lies between these limits which cover a greater range than the 95% limits.

6.2 CONFIDENCE LIMITS OF THE MEAN

Example 18. Calculation of standard error and 95% confidence limits for a large sample $(n > 30)$

Counts as in example 3 and therefore $\bar{x} = 10 \cdot 125$, $s^2 = 7 \cdot 465$, and $n = 80$.

$$\text{Standard error of the mean} = \sqrt{\frac{s^2}{n}} = \sqrt{\frac{7 \cdot 465}{80}} = 0 \cdot 306$$

Therefore the estimate of the population mean with its standard error is

$$\bar{x} \pm \sqrt{\frac{s^2}{n}} = 10 \cdot 125 \pm 0 \cdot 306$$

As $n > 30$, the normal approximation can be used and therefore the 95% confidence limits are

$$\bar{x} - 2\sqrt{\frac{s^2}{n}} \quad \text{to} \quad \bar{x} + 2\sqrt{\frac{s^2}{n}}$$

$$= 10 \cdot 125 - 2(0 \cdot 306) \text{ to } 10 \cdot 125 + 2(0 \cdot 306)$$

$$= 9 \cdot 513 \text{ to } 10 \cdot 737 \quad \text{or} \quad 10 \cdot 125 \pm 0 \cdot 612$$

Therefore the odds are about '19 to 1' that the true population mean (μ) lies between 9·51 and 10·74 nymphs in 100 cm². These confidence limits can be used to determine corresponding limits for an estimate of the total population in the sampling area. A preliminary estimate of 30,375 nymphs was made in section 6.1, and it is now possible to calculate the limits of this estimate. As there are between 9·513 and 10·737 nymphs in 100 cm², then there are between

$$\left(\frac{300,000}{100}\right) 9 \cdot 513 = 28{,}539 \quad \text{and} \quad \left(\frac{300,000}{100}\right) 10 \cdot 737 = 32{,}211$$

nymphs in 30 m². Therefore it is now possible to state that the total numbers in 30 m² of stream probably lies ($P = 0.95$) between 28,539 and 32,211 nymphs. A narrower estimate can only be obtained by increasing the number of sampling units in the sample, and it is misleading to give only a single value derived from the sample mean e.g. 30,375.

6.2.2 Small samples $(n < 30)$ from a Poisson series
(Random distribution)

The normal approximation cannot be applied to most small samples $(n < 30)$, but an exception is the Poisson series when the product nm is greater than 30 (m is the estimate of the Poisson parameter λ). As $\bar{x} = m = s^2$ in the Poisson series (see section 3.2.2), $\bar{x}$ is the best estimate of both m and s^2, and therefore the estimate of m (and hence λ) with its standard error is:

$$\bar{x} \pm \sqrt{\frac{\bar{x}}{n}}$$

Confidence limits for the estimate of the population mean μ ($\mu = \lambda = \sigma^2$) are therefore:

$$\bar{x} - t\sqrt{\frac{\bar{x}}{n}} \quad \text{to} \quad \bar{x} + t\sqrt{\frac{\bar{x}}{n}}$$

where t is approximately 2, or may be found in the column headed $2Q = 0.05$ (for 95% limits) of Student's t-distribution (Pearson & Hartley 1966, Table 12).

Published tables of confidence limits of a Poisson variable can be used when nm is less than 30, or when confidence limits are required for single counts which are assumed to come from a Poisson series. In Table 40 (Pearson & Hartley 1966), the 95% confidence limits are given in the column headed $1 - 2\alpha = 0.95$ (or $\alpha = 0.025$), and c is either the statistic m (estimated by $\bar{x}$) or a single count from a Poisson series. Alternatively, the 95% limits can be read off directly from Fig. 19 when $c < 30$. For values of c up to 300, there is the longer table of Crow & Gardner (1959).

When single counts from a Poisson series are used to estimate the population mean, the accuracy of the estimate depends upon the size of the count; e.g. the 95% confidence limits are $\pm 50\%$ (8 to 24) for a count of 16, $\pm 20\%$ (80 to 120) for a count of 100, and $\pm 10\%$ (360 to 440) for a count of 400. Therefore, if the Poisson series is a suitable model, the accuracy of the estimated population mean is improved by counting more invertebrates. The larger count may be from a larger sampling unit or from a combination of several small sampling units.

Example 19. **Calculation of 95% confidence limits for a small sample ($n < 30$) from a Poisson series**

Counts as in example 1, and therefore $m = \bar{x} = 11.273$, $s^2 = 7.415$, and $n = 11$. Agreement with a Poisson series was accepted ($P > 0.05$) and therefore the dispersion of the population was possibly random (see example 8A). As the product nm is greater than 30, the best estimate of m (and hence λ) with its standard error is:

$$\bar{x} \pm \sqrt{\frac{\bar{x}}{n}} = 11.273 \pm 1.012$$

Value of t (from Pearson & Hartley 1966, Table 12) = 2.23 for $v = n - 1 = 10$ degrees of freedom and $2Q = 0.05$ (for 95% limits). Therefore confidence limits for the estimate of the population mean μ ($= \lambda$) are:

$$\bar{x} - t\sqrt{\frac{\bar{x}}{n}} \quad \text{to} \quad \bar{x} + t\sqrt{\frac{\bar{x}}{n}}$$

$= 11.273 - 2.23(1.012)$ to $11.273 + 2.23(1.012)$
$= 9.016$ to 13.530
or 11.273 ± 2.257

Therefore the true population mean (μ) probably lies ($P = 0.95$) between 9.02 and 13.53 nymphs per sampling unit (500 cm²). These limits can be used to estimate the population in a section of stream, e.g. there are between 5,412 and 8,118 nymphs in 30 m² of stream.

6.2 CONFIDENCE LIMITS OF THE MEAN

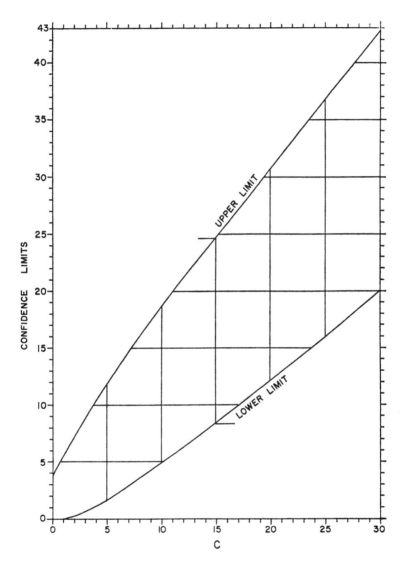

FIGURE 19. 95% confidence limits for c from a Poisson series. c is a single count or the mean ($\bar{x} = m$) of a small sample with $nm < 30$.

EXAMPLE: for count of 15, the confidence limits are: lower 8·5 and upper 24·5. Therefore the odds are 19 to 1 that the population mean lies between 8·5 and 24·5 invertebrates per sampling unit.

6.2.3 Small samples ($n < 30$) from a positive binomial distribution (Regular distribution)

The normal approximation can still be used if p (the probability of any given point in the sampling unit being occupied by an individual) and k (the maximum number of individuals a sampling unit could contain) are either of the following combinations. For k between 10 and 30, p should be between 0·4 and 0·6; whereas for k greater than 30, p can lie between 0·1 and 0·9. The normal approximation will tend to over-emphasise significance in the right hand tail and to under-emphasise significance in the left hand tail when $p > 0.5$, whilst the converse will be true for $p < 0.5$.

It is difficult to calculate confidence limits for small samples when $k < 10$, or when k is 10 to 30 with $p < 0.4$ or > 0.6. Approximate confidence limits can be quickly derived from 95% confidence limits for p, and these are given in extensive tables (Mainland, Herrera & Sutcliffe 1956; Documenta Geigy 1962) or charts (Pearson & Hartley 1966, Table 41). In these charts, values of k are printed along the curves, values of p are given on the abscissae (as c/n), and confidence limits are read off the ordinates. The lower abscissa and left-hand ordinate are used when $p < 0.5$, whereas the upper abscissa and right-hand ordinate are used when $p > 0.5$. As $p = \bar{x}/k$, the limits for the mean are obtained by multiplying the limits for p by k.

For example, in a small sample $\bar{x} = 14.4$, the variance is significantly smaller than the mean, and the estimated value of k is 20. Therefore $p = \bar{x}/k = 0.72$ and the 95% confidence limits for p are read off the right-hand ordinate as 0·48 and 0·89. Corresponding limits for the mean are therefore $0.48k$ and $0.89k = 9.6$ and 17.8. Therefore the odds are 19 to 1 ($P = 0.95$) that the true population mean (μ) lies between 9·6 and 17·8 animals per sampling unit.

As a regular distribution will rarely describe the dispersion of a Population (section 4.2), the above methods will rarely be needed. Very approximate confidence limits are rapidly obtained from limits for a Poisson variable (Fig. 19), but the latter limits are always wider than the corresponding limits for a positive binomial; e.g. the 95% limits for a mean of 14 are 8 to 24 (from Fig. 19) and are much wider than those of 9·5 and 17·8.

6.2.4 Small samples ($n < 30$) from a contagious distribution

The methods associated with the normal distribution cannot be applied directly to small samples from a contagious distribution ($s^2 > \bar{x}$), but can be applied to transformed counts (see section 3.2.4).

6.2 CONFIDENCE LIMITS OF THE MEAN

As the dispersion of a Population is frequently contagious (section 5), transformations are often necessary. Suitable transformations are obtained from the negative binomial, Taylor's power law, or logarithms.

(1) *Negative binomial.* The following methods are applicable when the negative binomial is successfully fitted to a sample. The choice of a suitable transformation depends upon the values of $\bar{x}$ and $\hat{k}$ (see Table 4, and method 3 for estimating k on page 56). Each count x in the sample is replaced by y and:

$$y = \log\left(x + \frac{k}{2}\right)$$

in the simplest transformation. Therefore the mean of the transformed counts ($\bar{y}$) is given by:

$$\bar{y} = \frac{\Sigma \log(x + k/2)}{n}$$

where n = number of sampling units = number of counts. As the distribution of the transformed counts is approximately normal and their expected variance is 0·1886 trigamma k (see Table 8), then the 95% confidence limits for $\bar{y}$ are:

$$\bar{y} \pm t \sqrt{\frac{0 \cdot 1886 \text{ trigamma } k}{n}}$$

where t is found in Student's distribution (Pearson & Hartley 1966, Table 12). These limits are transformed back to the original scale to give confidence limits for the population mean thus:

$$\text{antilog}\left(\bar{y} \pm t \sqrt{\frac{0 \cdot 1886 \text{ trigamma } k}{n}}\right) - \frac{k}{2}$$

The derived mean from the transformed counts, *i.e.* (antilog $\bar{y}$) $- k/2$, will always be smaller than the arithmetic mean ($\bar{x}$) of the original counts, except when $n = 1$ [*i.e.* for a single count, $\bar{x}$ = (antilog $\bar{y}$) $- k/2$]. Holmes & Widrig (1956) used this transformation to construct charts of 95% confidence limits for single counts with $k = 2$ (equivalent to $\lambda^2 = 0.5$ in charts) and $k = 5$ ($\lambda^2 = 0.2$). These limits are so wide that they are almost valueless: *e.g.* for a single count of 20 invertebrates with $k = 2$, the 95% limits $\simeq$ 3 to 100; and for count of 40, 95% limits $\simeq$ 7 to 200. Therefore very small samples ($n \simeq 1$) from a negative binomial will only give a rough estimate of the population mean and larger samples ($n \geqslant 10$) are necessary if a more accurate estimate is required, *i.e.* the confidence limits rapidly narrow as n increases (see example 20).

If the more complex $\sinh^{-1}$ transformation is used (see page 56), the 95% confidence limits for $\bar{y}$ are:

$$\bar{y} \pm t \sqrt{\frac{0.25 \text{ trigamma } k}{n}}$$

These limits have to be finally transformed back to the original scale by reversing the calculation of the original transformation. The greater accuracy of this complex transformation rarely justifies the laborious calculations and the simpler transformation, $\log(x+k/2)$, usually provides approximate limits, even with small means. A Poisson approximation is often applicable, when $\bar{x}$ is small (< 5) and k is fairly large (> 5). The logarithmic transformation, $\log(x+1)$, can be used if k is close to 2.

Example 20. Calculation of 95% confidence limits for a small sample from a negative binomial

The counts of a small sample ($n = 5$) are transformed to $\log(x+k/2)$ and the mean ($\bar{y}$) of the transformed counts $= 1.3522$, with $k = 5$.

0.1886 trigamma $k = 0.0417$ (from Table 8).

Value of t (from Pearson & Hartley 1966, Table 12) $= 2.776$ for $v = n-1 = 4$ degrees of freedom and $2Q = 0.05$. Therefore 95% confidence limits for the estimate of the population mean μ are:

$$\text{antilog}\left(\bar{y} \pm t \sqrt{\frac{0.1886 \text{ trigamma } k}{n}}\right) - \frac{k}{2}$$

$$= \text{antilog}\left(1.3522 \pm 2.776 \sqrt{\frac{0.0417}{5}}\right) - \frac{5}{2}$$

$$= [\text{antilog}(1.3522 \pm 0.2535)] - 2.5$$

$$= 10.05 \text{ to } 37.84 \text{ for a derived mean of } 20$$

Although these confidence limits are wide, they narrow rapidly with a further increase in sample size, *e.g.* check that the 95% confidence limits are 13.6 to 29.0 when $n = 10$, and 14.8 to 26.7 when $n = 15$.

(2) *Taylor's power law.* If the power law is suitable for a series of samples, then the appropriate variance-stabilising transformation is $y = x^p$; where y is the transformed count, x is the original count, and $p = 1 - b/2$ (see section 5.5). This transformation is applied when the negative binomial cannot be fitted to samples, or when a common transformation is required for comparisons of samples. A common

6.2 CONFIDENCE LIMITS OF THE MEAN

transformation is also obtained from a common k (section 5.2.3). When the power law and a common k are both applicable to the same series of samples (see examples 16, 17), the transformations derived from the power law are often easier to apply. The logarithmic transformation is used when $b = 2$ and $p = 0$.

Example 21. Calculation of 95% confidence limits for a small sample when the power law is applicable

In example 17, the power law was successfully fitted to a series of 10 samples. The estimate of the parameter b was 1·57 and therefore $p=0·2$, $y=x^{0·2}$. This transformation will be applied to the second sample of the series (see Table 10 and example 15). The 10 counts of the sample are:

x	4	5	8	14	14	15	15	19	28	36
y	1·32	1·38	1·52	1·70	1·70	1·72	1·72	1·80	1·95	2·05

The transformed counts (y) are either obtained directly from Table 1 of Healy & Taylor (1962), or calculated thus:
e.g., for first count of 4,

$$y = x^{0·2} = 4^{0·2} = \text{antilog}\,(0·2 \log 4)$$
$$= \text{antilog}\,[0·2(0·602)]$$
$$= \text{antilog}\,0·1204 = 1·320$$

The mean of the transformed counts $= \bar{y} = 1·6860$.

Value of t (from Pearson & Hartley 1966, Table 12) = 2·262 for $v = n-1 = 9$ degrees of freedom and $2Q = 0·05$.

$$\text{Standard error of } \bar{y} = \sqrt{\frac{\text{variance of transformed counts}}{n}}$$
$$= 0·0725$$

Therefore 95% confidence limits for $\bar{y}$ are:

$$= \bar{y} \pm t \text{ (standard error of } \bar{y})$$
$$= 1·686 \pm 2·262\,(0·0725)$$
$$= 1·522 \quad \text{to} \quad 1·850$$

These limits and $\bar{y}$ have to be transformed back to the original scale to give a derived mean and confidence limits, e.g.:

$$\text{derived mean} = \text{antilog}\left(\frac{\log \bar{y}}{0·2}\right) = \text{antilog}\left(\frac{0·22687}{0·2}\right) = 13·63$$

Lower 95% confidence limit

$$= \text{antilog}\left(\frac{\log 1·522}{0·2}\right) = 8·17$$

Upper 95% confidence limit

$$= \text{antilog}\left(\frac{\log 1·850}{0·2}\right) = 21·67$$

Therefore the true population mean (μ) probably lies ($P=0.95$) between 8·17 and 21·67 individuals per sampling unit. Note that the derived mean of 13·63 is lower than the arithmetic mean of the original counts ($\bar{x}=15.80$ in example 15).

Confidence limits can also be calculated from a transformation derived from the negative binomial, namely $y = \log(x+k/2)$. The estimate of k is either the sample estimate of k ($\hat{k} = 3.2$ in example 15), or a common k when a common transformation is required ($k_c = 2.95$ in example 16).

(3) *Logarithmic transformation.* This transformation, or its variant $\log(x+1)$, is an accurate transformation when $k \simeq 2$ in the negative binomial, or when $b \simeq 2$ in the power law. It is also an approximate transformation for most small samples from a contagious distribution, and is used when other transformations cannot be applied, or when the samples are too small to justify a more accurate transformation. A logarithmic transformation: (1) is easy to calculate and apply, (2) usually satisfies the most important function of a transformation, namely the stabilisation of variance (section 3.2.4), and (3) enables confidence limits to be expressed in terms of a derived mean divided and multiplied by a common factor. Therefore a logarithmic transformation is the most useful transformation for small samples, but a more precise transformation is usually obtained from Taylor's power law. A straight logarithmic transformation is used for most small samples, but $\log(x+1)$ must be used when zero counts are present in the sample.

The mean of the transformed counts ($\bar{y}$) is given by:

$$\bar{y} = \frac{\Sigma \log x}{n}$$

where n = number of sampling units = number of counts. The 95% confidence limits for $\bar{y}$ are:

$$\bar{y} \pm t \sqrt{\frac{\text{variance of transformed counts}}{n}}$$

where t is found in Student's t-distribution (Pearson & Hartley 1966, Table 12). Antilogarithms of these limits give the 95% confidence limits for the population mean. The derived mean (antilog $\bar{y}$) from the transformed counts is equal to the geometric mean of the original counts, and is always lower than the arithmetic mean ($\bar{x}$) of the original counts. If comparisons are to be made with other arithmetic means, a small adjustment has to be made to the derived mean (see section 3.2.4).

When confidence limits are written after the derived mean, they must be given as antilogarithms, and the $\pm$ of the log scale becomes

6.2 CONFIDENCE LIMITS OF THE MEAN

$\overset{\times}{\div}$ on the arithmetic scale, e.g. if $\bar{y} = 2\cdot00$, standard error $= 0\cdot15$, and $t = 2$; then

$$\bar{y} \pm t \text{ (standard error of } \bar{y}) = 2\cdot00 \pm 0\cdot30$$

and the confidence limits of the derived mean are

$$100 \overset{\times}{\div} 2 = 50 \text{ to } 200 \text{ (not } 100 \pm 2)$$

Bagenal (1955) gives a clear account of the relationships between some parameters during and after a logarithmic transformation.

Example 22A. Use of the logarithmic transformation to calculate 95% confidence limits

If the counts of a small sample ($n=5$) are 98, 22, 72, 214, 67; then $\bar{x}=94\cdot60$, $s^2 = 5202\cdot80$ and clearly $s^2 > \bar{x}$ (see example 8B). The transformed counts ($\log x$) are 1·99123, 1·34242, 1·85733, 2·33041, 1·82607. Arithmetic mean of transformed counts $= \bar{y} = 1\cdot8695$.

Variance of transformed counts

$$= \frac{\Sigma(\log x - \bar{y})^2}{n-1} = 0\cdot1268$$

Value of $t = 2\cdot776$, for $n-1 = 4$ degrees of freedom and $2Q = 0\cdot05$ (for 95% limits).

Therefore 95% confidence limits for $\bar{y}$ are:

$$\bar{y} \pm t \sqrt{\frac{\text{variance of transformed counts}}{n}}$$

$$= 1\cdot8695 \pm 2\cdot776 \sqrt{\frac{0\cdot1268}{5}}$$

$$= 1\cdot8695 \pm 0\cdot4419$$

$$= 1\cdot4276 \quad \text{to} \quad 2\cdot3114.$$

Antilogs of these limits are 26·77 to 204·83. Antilogs $\bar{y} = 74\cdot05$, and antilog $\pm 0\cdot4419 = \overset{\times}{\div} 2\cdot77$.

Therefore the derived mean (= geometric mean) = 74·05 and the 95% confidence limits are $74\cdot05 \overset{\times}{\div} 2\cdot77 = 27$ to 205. Therefore the odds are 19 to 1 that the true population mean (μ) lies between 27 and 205 animals per sampling unit. These limits can now be used to determine corresponding limits for the total numbers in the sampling area (see example 18).

Example 22B. Use of the log ($x+1$) transformation to calculate 95% confidence limits

The following counts of *Lymnaea peregra* were obtained from a small sample ($n=4$):

$$0, 3, 9, 10 \text{ and } \bar{x} = 5\cdot5$$

As there is a zero count, the transformation $\log(x+1)$ is used, and the transformed counts are:

$$0\cdot0000, 0\cdot6021, 1\cdot0000, 1\cdot0414$$

Arithmetic mean of transformed counts = $\bar{y}$ = 0·6609. Variance of transformed counts = 0·2333. Value of t = 3·182, for $v = n-1 = 3$ degrees of freedom and $2Q = 0·05$ (for 95% limits).

Therefore 95% confidence limits for $\bar{y}$ are:

$$\bar{y} \pm t \sqrt{\frac{\text{variance of transformed counts}}{n}}$$

$$= 0·6609 \pm 3·182 \sqrt{\frac{0·2333}{4}}$$

$$= 0·6609 \pm 0·7685$$

$$= -0·1076 \text{ to } 1·4295 = \bar{1}·8924 \text{ to } 1·4294$$

Derived mean = (antilog $\bar{y}$) − 1 = 4·58 − 1 = 3·58.

This derived mean is always less than the arithmetic mean of the original counts ($\bar{x} = 5·5$).

95% confidence limits for this derived mean are:

Lower limit = (antilog $\bar{1}·8924$) − 1 = 0·78 − 1 ≃ 0
Upper limit = (antilog 1·4294) − 1 = 26·88 − 1 = 25·88.

The confidence limits can also be expressed as:

$$[\text{antilog}(0·6609 \pm 0·7685)] - 1$$
$$= (4·58 \overset{\times}{\div} 5·87) - 1$$
$$= 0 \text{ to } 26.$$

Therefore the true population mean (μ) probably lies ($P = 0·95$) between 0 and 26 individuals per sampling unit. These limits are very wide and can only be narrowed by increasing the number of sampling units (n).

6.3 Summary

A sample mean is usually used to estimate a population mean, and the latter is often used to estimate the total number of invertebrates in an area of bottom. The errors of these estimates are usually expressed as confidence limits, which state the range in which the population mean probably lies (usually $P = 0·95$). Limits for an estimate of the total numbers in an area of bottom are simply obtained by multiplying upper and lower confidence limits by

$$\frac{\text{area of bottom}}{\text{area of sampling unit}}$$

(see example 18).

Confidence limits for the estimate of the population mean are calculated by the following methods:

(A) Large samples ($n \geqslant 30$), use normal approximation:
$$\bar{x} \pm t\sqrt{s^2/n} \text{ (section 6.2.1)}$$

6.3 SUMMARY

(B) Small samples ($n < 30$):

(B.1) If $s^2 = \bar{x}$, use methods for Poisson (section 6.2.2):
when $nm > 30$, use $\bar{x} \pm t\sqrt{\bar{x}/n}$
when $nm < 30$, refer to Fig. 19 or Table 40 in Pearson & Hartley (1966).

(B.2) If $s^2 < \bar{x}$, use methods for positive binomial (section 6.2.3). These methods will rarely be needed.

(B.3) If the small sample is from a contagious distribution with $s^2 > \bar{x}$, then a transformation is always necessary (section 6.2.4).

(B.3.a) If the negative binomial is successfully fitted to a sample, $\log(x+k/2)$ is often a suitable transformation and confidence limits are given by:

$$\text{antilog}\left(\bar{y} \pm t\sqrt{\frac{0.1886 \text{ trigamma } k}{n}}\right) - \frac{k}{2}$$

(B.3.b) If the negative binomial cannot be fitted, a logarithmic transformation is often used and confidence limits are given by:

$$\text{antilog}\left(\bar{y} \pm t\sqrt{\frac{\text{variance of transformed counts}}{n}}\right)$$

The log transformation is probably the most useful transformation, and is used when samples are too small to justify a more accurate transformation. If zero counts are present, the transformation $\log(x+1)$ is used. The log transformation will be suitable for most purposes, but a more precise transformation can be obtained from Taylor's power law (see section 5.5 and 6.2.4).

VII COMPARISON OF SAMPLES

The means of two samples will rarely be the same, and several samples with different means may be all from the same population (see section 6.1). Therefore a statistical test should indicate whether a difference between sample means is significant, or whether this difference is covered by the error of the estimated population mean.

First we must make an assumption called the *null hypothesis* (H_0). This is simply a general description of what we expect to happen according to a standard hypothesis. The usual null hypothesis is that the samples are from the same population and therefore differences between sample means are within the accepted error of the population mean. When we compare two different sample means, we have to calculate the probability (P) of obtaining the same (or a larger) difference in means when the samples are from the same population (*i.e.* when H_0 is true). The null hypothesis is accepted when this probability (P) is fairly large, but H_0 is rejected when P is significantly small. A 5% level of significance is frequently used ($P = 0.05$), and H_0 is accepted when P is greater than 0.05. Therefore H_0 is rejected when P is less than 0.05 ($P < 0.05$), and we conclude that the sample means are significantly different at the 5% level. When $P = 0.05$, the odds are only 1 in 20 that the difference in means (or a larger difference) could have arisen by chance under H_0. If these odds are not acceptable, a more stringent level of significance is required for the rejection of H_0, *e.g.* odds of 1 in 100 ($P = 0.01$) or 1 in 1000 ($P = 0.001$). The different levels of significance are often indicated by asterisks thus:

     *   = $P < 0.05$ Significant
    ** = $P < 0.01$ Highly significant
   *** = $P < 0.001$ Very highly significant

The 5% level ($P = 0.05$) is suitable for most purposes, but it is always better to state the actual value of P attained in the test, rather than to simply record $P < 0.05$.

Therefore a significance test provides a reasonably objective basis for reaching decisions. Emphasis has been placed on differences between sample means, but the same theory can be applied to comparisons of other statistics, *e.g.* variance. Two types of error can occur in a statistical test. A *Type I error* is to reject H_0 when it is true. The probability of making this error decreases as the stated level of significance decreases, *e.g.* the probability of a Type I error is greater at $P = 0.05$ than at $P = 0.01$. A *Type II error* is to accept H_0 when it is false. It is obvious that a decrease in a Type I error will increase the probability of a Type II error for any given sample size. Both errors are reduced by an increase in sample size, *i.e.* an increase in the number (n) of sampling units. The power of a test is defined as the probability of rejecting H_0 when it is false, *i.e.* power = 1 − probability of a Type II error. The ideal statistical test has a small probability of rejecting H_0 when it is true, and a large probability of rejecting H_0 when it is false. Therefore a statistical test is not an infallible guide and can never *prove* a particular hypothesis. There is always a possibility that an alternative hypothesis exists.

Significance tests can be *parametric* or *non-parametric*. In the following parametric tests, the null hypothesis assumes that a known distribution is a suitable model for the samples (*e.g.* normal distribution, or Poisson series), and specifies that parameters of this distribution (*e.g.* μ, σ^2, λ) are the same for the samples under test. The non-parametric tests in section 7.2 do not require this assumption.

7.1 Parametric tests

The following parametric tests can be used when the distribution of the parent populations is known from previous experience or from theoretical considerations (*e.g.* central-limit theorem), or is deduced from the sample itself (*e.g.* Poisson series). When the conditions for a test are possibly fulfilled, they should be stated, *e.g.* "On condition that both samples originate from the same normally-distributed population, there is a significant difference between the sample means". When the conditions for a test are obviously not fulfilled, parametric tests should not be applied directly to the samples.

7.1.1 *Methods associated with the normal distribution, and their application to large samples ($n > 50$)*

These methods are parametric and require at least three assumptions (see section 3.2.4). Some of these methods can be applied to large samples ($n > 50$) when the central-limit theorem is applicable (see section 6.2.1). The methods should not be used indiscriminately and cannot be applied directly to most small samples ($n < 50$). This difficulty may be overcome by transforming the counts (sections 3.2.4; 6.2.4), or by using non-parametric methods. As the normal distribution and its associated methods form the major part of most statistical textbooks, only a brief outline of the more important methods is included here.

(1) *Comparing the means of two large samples* ($n > 50$). The null hypothesis is that both samples come from the same population and therefore have the same means and variances ($\mu_1 = \mu_2$ and $\sigma_1^2 = \sigma_2^2$). Arithmetic mean ($\bar{x}$), variance (s^2), and number of sampling units (n) are $\bar{x}_1$, s_1^2, and n_1 for the first sample; and $\bar{x}_2$, s_2^2, and n_2 for the second sample. As the samples are large ($n > 50$), the central-limit theorem is applicable. Therefore the parent distribution of each sample mean is assumed to be approximately normal, and the standard errors of the means are $\sqrt{s_1^2/n_1}$ and $\sqrt{s_2^2/n_2}$ (see section 6.1). Under the null hypothesis, the difference between the means is zero ($\mu_1 - \mu_2 = 0$). As the central-limit theorem is applicable to each sample, the difference in observed means has a theoretically normal distribution about zero and a standard deviation equal to the standard deviation of difference in means. Therefore the actual difference in sample means ($\bar{x}_1 - \bar{x}_2$) is d standard deviations away from zero, where:

$$d = \frac{\text{difference in sample means}}{\text{standard deviation of difference}} = \frac{\bar{x}_1 - \bar{x}_2}{\sqrt{\dfrac{s_1^2}{n_1} + \dfrac{s_2^2}{n_2}}}$$

Therefore d is a normal variable with zero mean and unit standard deviation. The ratio d is either positive or negative, and is significant at the 5% level ($P = 0.05$) when the absolute value of d (*i.e.* without regard for sign) is greater than 1·96. The difference in sample means is significant at the 1% level ($P = 0.01$) when d is greater than 2·58, and at the 0·1% level ($P = 0.001$) when $d > 3.29$. This method can be applied to large samples ($n > 50$) from random, regular, or contagious distributions.

As the population variance is estimated from the same sample as the mean (σ^2 estimated by s^2), it is strictly correct to use t instead of d in the above formula and to refer t to tables of Student's t-distribution (see section 7.1.3). With large samples, there is little difference between the values of d and t, *e.g.* at 5% significance level $d = 1.96$ and $t = 1.98$ with 98 degrees of freedom (for $n_1 = n_2 = 50$).

Although the above test is used to compare sample means, it is important to realise that a significant difference between variances, or both means and variances, will also give a significant value of d (or t). Therefore the F-test should be used to check if the variances are significantly different.

Example 23. Comparison of means ($n > 50$)

Statistics of two samples are:

$\bar{x}_1 = 10.125$, $s_1^2 = 7.465$, and $n_1 = 80$ (from example 3)
$\bar{x}_2 = 12.245$, $s_2^2 = 8.855$, and $n_2 = 60$

7.1 PARAMETRIC TESTS

The null hypothesis is that both samples come from the same population and therefore have the same means and variances. The variances are not significantly different (see example 24). As the samples are large, the central-limit theorem is applicable.

$$d = \frac{\bar{x}_1 - \bar{x}_2}{\sqrt{\dfrac{s_1^2}{n_1} + \dfrac{s_2^2}{n_2}}} = \frac{10\cdot125 - 12\cdot245}{\sqrt{\dfrac{7\cdot465}{80} + \dfrac{8\cdot855}{60}}}$$

$$= \frac{-2\cdot12}{\sqrt{0\cdot24089}} = -4\cdot3194$$

As the absolute value of d is greater than 3·29, the difference in sample means is significant at the 0·1% level ($P < 0\cdot001$), i.e. difference is very highly significant (cf. example 25).

(2) *Comparing the variances of two large samples* ($n > 50$). The null hypothesis (H_0) is that both samples come from the same normally-distributed population, and therefore have the same variances ($\sigma_1^2 = \sigma_2^2$). The two variance estimates are s_1^2 and s_2^2 and the equality of these estimates is tested by the "variance-ratio" or "*F*-test".

$$F = \frac{s_1^2}{s_2^2}$$

where s_1^2 is always the largest of the two variances, and therefore F is never less than unity. The appropriate value of F for a chosen level of significance is found in tables (Pearson & Hartley 1966, Table 18), with $v_1 = n_1 - 1$ degrees of freedom in the numerator and $v_2 = n_2 - 1$ degrees of freedom in the denominator. H_0 is accepted when this tabulated value is larger than the calculated value. If the tabulated value is exceeded by the calculated value, the difference in variances is significant at the chosen level of significance. The tables are usually used in the Analysis of Variance and are therefore one-tailed. As a two-tailed test is required here, the table of upper 2·5% points is used for the 5% level of significance ($P = 0\cdot05$), and the table of upper 0·5% points is used for the 1% level of significance ($P = 0\cdot01$).

Example 24. Comparison of variances ($n > 50$)
Variances from example 23 and therefore $s_1^2 = 8\cdot855$ (s_1^2 is always largest variance), $s_2^2 = 7\cdot465$, $n_1 = 60$, $n_2 = 80$.

$$F = \frac{s_1^2}{s_2^2} = \frac{8\cdot855}{7\cdot465} = 1\cdot1862$$

Degrees of freedom, $v_1 = n_1 - 1 = 59$, $v_2 = n_2 - 1 = 79$.

The nearest value of F from tables (Pearson & Hartley 1966, Table 18, upper 2·5% points) is 1·67 for $v_1 = 60$ and $v_2 = 60$, or 1·53 for $v_1 = 60$ and $v_2 = 120$. As the calculated value of F is well below the tabulated values, the variances are not significantly different at the 5% level ($P > 0·05$).

(3) *Coefficient of variation* (C). This coefficient is sometimes used to compare the relative variability of samples. It is the term applied to the standard deviation when it is expressed as a percentage of the sample mean.

$$C = s\left(\frac{100}{\bar{x}}\right)$$

where $\bar{x}$ = sample mean, and s = standard deviation of sample.

(4) *Analysis of variance and the comparison of more than two samples*. Analysis of variance is one of the most powerful and useful techniques in statistics. The total variation in a set of data is partitioned into components associated with possible sources of variability. Then the relative importance of the different sources is assessed by F-tests between each component of variation and the "error" variation. This is a parametric technique which requires the following assumptions:

(1) The samples are independently drawn from normally-distributed populations.

(2) The parent populations all have the same variance, and therefore mean and variances are independent.

(3) The components of variance must be additive.

As these conditions are never fulfilled by counts from bottom samples, transformations are always necessary before the analysis of variance is applied (see sections 3.2.4, 6.2.4).

The analysis of variance is fully described in most textbooks (*e.g.* Bailey 1959, Campbell 1967, Moroney 1956), and a detailed account is given in Snedecor & Cochran (1967; Chapters 10, 11, 12). As the basic principles are often difficult to understand, they are briefly described in this section.

The two common designs for the analysis are:

(1) One-way analysis of variance (or completely randomised design) when comparisons are made between a number of independent random samples, one sample from each population. The counts are only classified in one direction and the number of counts in each sample can be different.

(2) Two-way analysis of variance (or randomised block design) when the counts of related samples are matched in groups, and comparisons are made between groups and between samples. The counts are thus classified in two directions. This analysis can be extended to compare further groupings of the same counts.

The one-way analysis is the easiest to understand and the transformed counts of each sample are first arranged in a table thus:

	Counts	Number of counts	Total	Mean
Sample 1	$x_{11}\ x_{12}\ x_{13}\ ...\ x_{1n}$	n_1	T_1	$\bar{x}_1$
Sample 2	$x_{21}\ x_{22}\ x_{23}\ ...\ x_{2n}$	n_2	T_2	$\bar{x}_2$
Sample i	$x_{i1}\ x_{i2}\ x_{i3}\ ...\ x_{in}$	n_i	T_i	$\bar{x}_i$
Totals		$N = \Sigma n_i$	$\Sigma T_i = \Sigma x$	$\bar{x} = \dfrac{\Sigma x}{N}$

where i = number of samples, N = total number of sampling units = total number of counts, $\bar{x}$ = overall mean for all counts in all samples. $\Sigma x = \Sigma T_i$ = total number of individuals in all counts. In each sample, x is a particular count, and the two subscripts indicate the sample number and the count number.

Different *sums of squares* are now calculated [sum of squares is the numerator in the formula for variance, *i.e.* $\Sigma(x-\bar{x})^2$].

(1) Total sum of squares about the overall mean ($\bar{x}$)

$$S_T = \Sigma(x-\bar{x})^2 = \Sigma(x^2) - \frac{(\Sigma x)^2}{N}$$

(2) Sum of squares between samples

$$S_2 = \Sigma\left(\frac{T_i^2}{n_i}\right) - \frac{(\Sigma x)^2}{N} = \left[\frac{T_1^2}{n_1} + \frac{T_2^2}{n_2} + \cdots \frac{T_i^2}{n_i}\right] - \frac{(\Sigma x)^2}{N}$$

(3) Sum of squares of the residual variation, *i.e.* variation between sampling units within the samples

$$S_1 = S_T - S_2 \quad \text{or}$$
$$S_1 = \Sigma \text{ (sum of squares for each sample)}$$
$$ \text{all}$$
$$ \text{samples}$$

$$= \Sigma\left[\Sigma(x_i^2) - \frac{(T_i^2)}{n_i}\right]$$

The total sum of squares is thus separated into two components. One corresponds to the variation between sample means (S_2) and the other is the so-called *residual* or *error* sum of squares (S_1). The latter component arises from sampling errors and variation inherent in a sample, *e.g.* variation due to a contagious dispersion. The total degrees of freedom $(N-1)$ is also divided between the components, and the mean square (or variance) is then calculated from the following formulae:

Source of variation	Sums of squares	Degrees of freedom	Mean square (variance)
Between samples	S_2	$i-1$	$s_2^2 = \dfrac{S_2}{i-1}$
Between sampling units within the samples (residual variance)	S_1	$N-i$	$s_1^2 = \dfrac{S_1}{N-i}$
Total	S_T	$N-1$	

s_2^2 represents the dispersion of the sample means ($\bar{x}_i$) around the overall mean ($\bar{x}$), s_1^2 represents the dispersion of the individual counts around their respective sample means.

The null hypothesis (H_0) is that all samples come from normally-distributed populations with the same means ($\mu_1 = \mu_2 = \cdots = \mu_i$) and variances. Therefore s_1^2 and s_2^2 are estimates of the same variance. This hypothesis is tested by the F-test.

$$F = \frac{s_2^2}{s_1^2}$$

with $v_1 = i-1$ and $v_2 = N-i$ degrees of freedom in the numerator and denominator respectively. H_0 is accepted when F is less than the upper 5% point ($P = 0.05$) in published tables, and H_0 is rejected when F is larger than the tabulated value (Pearson & Hartley 1966, Table 18).

Confidence limits for each sample mean (*e.g.* $\bar{x}_1, \bar{x}_2, \ldots \bar{x}_i$) are given by:

$$\bar{x}_i \pm t \sqrt{\frac{s_1^2}{n_i}}$$

where s_1^2 is the residual variance for the set of samples, n_i = number of counts in the sample, and t is found in Student's distribution. These limits are for the mean of the transformed counts in each sample, and their calculation is explained in section 6.2.4.

The calculations in a two-way analysis are similar to those in the one-way analysis. Each count is now classified in two directions, and the transformed counts of each sample are arranged in a table thus:

7.1 PARAMETRIC TESTS

	Groups 1 2 3 n	Sample totals	Sample means
Sample 1	$x_{11}\ x_{12}\ x_{13}\ ...\ x_{1n}$	T_1	$\bar{x}_1$
Sample 2	$x_{21}\ x_{22}\ x_{23}\ ...\ x_{2n}$	T_2	$\bar{x}_2$
Sample i	$x_{i1}\ x_{i2}\ x_{i3}\ ...\ x_{in}$	T_i	$\bar{x}_i$
Group totals	$B_1\ B_2\ B_3\ ...\ B_n$	Σx	$\bar{x} = \dfrac{\Sigma x}{N}$

where the number of counts per sample (n) is the same for all samples, and B_n is the total number of individuals in each group. The groups could represent different bottom types (*e.g.* different ranges of stone size or water velocity, or different categories of vegetation), or the two-way classification could be used to compare different sources of variation (*e.g.* different months against different stations).

The calculations for the total sum of squares and sum of squares between samples are the same as those for the one-way analysis. The sum of squares between groups is given by:

$$S_3 = \sum \left(\frac{B_n^2}{i}\right) - \frac{(\Sigma x)^2}{N} = \left[\frac{B_1^2}{i} + \frac{B_2^2}{i} + \cdots \frac{B_n^2}{i}\right] - \frac{(\Sigma x)^2}{N}$$

The residual sum of squares is found by subtraction thus:

$$S_1 = S_T - S_2 - S_3$$

The calculations for degrees of freedom and mean squares are summarised below:

Source of variation	Sum of squares	Degrees of freedom	Mean square (variance)
Between groups	S_3	$n-1$	$s_3^2 = \dfrac{S_3}{n-1}$
Between samples	S_2	$i-1$	$s_2^2 = \dfrac{S_2}{i-1}$
Residual variance	S_1	$(n-1)(i-1)$	$s_1^2 = \dfrac{S_1}{(n-1)(i-1)}$
Total	S_T	$ni-1$ $= N-1$	

Confidence limits for each sample mean are given by:

$$\bar{x}_i \pm t \sqrt{\frac{s_1^2}{n}}$$

Both these methods of analysis can be extended to comparisons of several sources of variation, *e.g.* between months, between stations, between samples within stations, etc. It is also possible to calculate the *interaction* between different sources of variation. A full description of these methods is given in advanced textbooks (*e.g.* Snedecor & Cochran 1967). As the calculations in the analysis of variance are laborious, the technique is usually applied to the transformed counts of small samples (see examples 27 and 28).

(5) *Correlation between two variables.* Variations in the numbers of a species may be related to variations in the numbers of another species, or to variations in an environmental factor. The *product-moment correlation coefficient* (r) is frequently used to measure the degree of correlation between two variables. Its calculation is fully described in most textbooks (*e.g.* Bailey 1959, Snedecor & Cochran 1967) and only a brief summary is included here.

$$r = \frac{\Sigma(x-\bar{x})(y-\bar{y})}{\sqrt{[\Sigma(x-\bar{x})^2 \Sigma(y-\bar{y})^2]}}$$

where x is one variable with mean $\bar{x}$, and y is the other variable with mean $\bar{y}$. There is a value of y for each x, and the total number of pairs $= n$. The correlation is either positive ($+r$) or negative ($-r$), and is significant at the 5% level ($P = 0.05$) when r exceeds the tabulated value for $v = n-2$ degrees of freedom and $2Q = 0.05$ (Pearson & Hartley 1966, Table 13).

Use of this coefficient requires a *bivariate normal distribution*, *i.e. both* variables in the comparison must be normally distributed. As this condition is never fulfilled by the counts of a species, at least one transformation is always necessary. If the other variable is an environmental factor, the measurements may be normally distributed, *e.g.* measurements of stone size, temperature, water velocity.

A significant correlation must be interpreted with caution. The correlation may not be due to the direct influence of one variable on the other, but to the influence of unknown factors on both variables. "*Post hoc, ergo propter hoc*" (or "*cum hoc propter hoc*").

Regression analysis also examines the relationship between two variables and provides an equation relating one variable to another. In the calculation of r, no distinction was made between the two variables, whereas a regression line describes the average change in a dependent variable (y) for a unit change in an independent variable (x). The dependent variable is usually the numbers of a species and the independent variable is often an environmental factor. The relationship between the two variables is therefore the regression of y

on x, and the regression of x on y is often meaningless; *e.g.* regression of numbers (y) on stone size (x) is possible, but x on y implies that changes in numbers can affect the stone size! The calculations for a linear regression are given in example 17B, and are fully described in most textbooks (*e.g.* Bailey 1959, Snedecor & Cochran 1967).

A full discussion of the model in linear regression is beyond the scope of this booklet. One important assumption of this model is that the conditional distribution of y given x is normal (see Snedecor & Cochran, section 6.4). Therefore the dependent variable (y) must be transformed when it is the numbers of a species. No assumptions are made about the distributional properties of the independent variable (x), but there must be no appreciable experimental error in the measurement of x. If the regression line is curved, both variables can be transformed to obtain a linear regression, *e.g.* use of double-log scale in Taylor's power law. The dependence of one variable on two or more independent variables can be investigated by the more complex method of multiple regression analysis (Snedecor & Cochran 1967, Chapter 13).

7.1.2 *Samples from a Poisson series (random distribution)*

The normal deviate (d) is also used to compare the means of two samples from Poisson series, when the product nm is greater than 30 for each sample ($m = \bar{x} = s^2$, and m is the estimate of the Poisson parameter λ). $\bar{x}$ is the best estimate of both m and s^2 in the Poisson series (see section 3.2.2), and therefore:

$$d = \frac{m_1 - m_2}{\sqrt{\dfrac{m_1}{n_1} + \dfrac{m_2}{n_2}}} = \frac{\bar{x}_1 - \bar{x}_2}{\sqrt{\dfrac{\bar{x}_1}{n_1} + \dfrac{\bar{x}_2}{n_2}}}$$

The estimate (m) of the Poisson parameter λ, the arithmetic mean ($\bar{x}$), and the number of sampling units (n) are m_1, $\bar{x}_1$, and n_1 for the first sample, and m_2, $\bar{x}_2$, and n_2 for the second sample. The products $n_1 m_1$ and $n_2 m_2$ must be both greater than 30.

The null hypothesis (H_0) is that both samples come from the same population, and therefore the two Poisson parameters are equal ($\lambda_1 = \lambda_2$).

When the absolute value of d is greater than 1·96, H_0 is rejected at the 5% level ($P = 0.05$), *i.e.* the difference in sample means is significant at the 5% level (see also section 7.1.1, method 1).

Example 25. Comparison of means of two samples from Poisson series ($nm > 30$)
Statistics of two samples are:

$\bar{x}_1 = m_1 = 11 \cdot 273$, and $n_1 = 11$ (from example 1)
$\bar{x}_2 = m_2 = 8 \cdot 600$, and $n_2 = 5$.

Agreement with a Poisson series was accepted for both samples ($P > 0 \cdot 05$), and both $n_1 m_1$ and $n_2 m_2$ are greater than 30.

$$d = \frac{\bar{x}_1 - \bar{x}_2}{\sqrt{\frac{\bar{x}_1}{n_1} + \frac{\bar{x}_2}{n_2}}} = \frac{11 \cdot 273 - 8 \cdot 600}{\sqrt{\frac{11 \cdot 273}{11} + \frac{8 \cdot 6}{5}}} = +1 \cdot 6134$$

As the absolute value of d is less than $1 \cdot 96$, the difference in sample means is not significant at the 5% level ($P > 0 \cdot 05$). Therefore H_0 is accepted at the 5% level, and it is concluded that the two Poisson parameters are equal ($\lambda_1 = \lambda_2$) (cf. example 23).

Table 36 in Pearson & Hartley (1966) is used when nm is less than 30 for one or both samples, or when comparisons are made between single counts from Poisson series. The two Poisson variables are denoted by a and b ($b < a$), and the sum of the two counts ($r = a+b$) is given in the left-hand column of Table 36. There is a significant difference between the two counts when b is less than, or equal to, the tabulated value for a given level of probability. Probability levels must be doubled for a double-tailed test, e.g. column headed $0 \cdot 025$ is used for a test at the 5% level of significance, and $0 \cdot 005 \equiv 1\%$ level. For example, means of 12 and 3 are obtained from two samples from Poisson series ($nm < 30$). Therefore $a = 12$, $b = 3$, and $r = 15$. As b is equal to the tabulated value for $r = 15$ and $P = 0 \cdot 025$, the means are significantly different at the 5% level, but not at the 1% level (tabulated value $= 2$, for $r = 15$ and $P = 0 \cdot 005$).

As the variance and mean are equal in a Poisson series, the F-test is not used for a direct comparison of variances. The coefficient of variation is given by:

$$C = 100 \left(\frac{s}{\bar{x}}\right) = 100 \left(\frac{\sqrt{\bar{x}}}{\bar{x}}\right) = \frac{100}{\sqrt{\bar{x}}}$$

The other methods associated with the normal distribution cannot be applied directly to counts from Poisson series, but can be applied to transformed counts. Therefore each count must be replaced by its square root (or $\sqrt{x+0 \cdot 05}$ if some counts are less than 10), before analyses of variance, correlation coefficients, or regression analyses are applied (see section 3.2.4 and Table 4).

7.1.3 Small samples ($n < 50$) from contagious distributions

The methods associated with the normal distribution cannot be applied directly to small samples from contagious distributions, but all except the normal deviate test (method 1) can be applied to transformed counts. Suitable transformations are obtained from the negative binomial, Taylor's power law, or logarithms. The application and choice of a suitable transformation are discussed in section 6.2.4. It is not possible to give examples for all possible combinations of the three transformations and the different methods associated with the normal distribution. Therefore the examples are restricted to the F-test, t-test and analysis of variance on counts transformed by means of the power law (example 26), and to analyses of variance on counts transformed to logarithms (examples 27, 28).

Student's t-test (example 26A) or a one-way analysis of variance (example 26B) can be used to compare two small samples. As the t-test is fully described in statistical textbooks (*e.g.* Bailey 1959), only a brief account is given here. The null hypothesis (H_0) is that the two samples are drawn from populations with the same means and variances ($\mu_1 = \mu_2$, $\sigma_1^2 = \sigma_2^2$). For small samples ($n < 50$), the populations must also be normal, whereas for large samples ($n > 50$) the central-limit theorem is applicable (see section 7.1.1, method 1).

$$t = \frac{\text{difference between means}}{\text{standard error of difference}} = \frac{\bar{x}_1 - \bar{x}_2}{\sqrt{s^2 \left(\frac{1}{n_1} + \frac{1}{n_2}\right)}}$$

$$\text{where } s^2 = \frac{\Sigma(x_1 - \bar{x}_1)^2 + \Sigma(x_2 - \bar{x}_2)^2}{n_1 + n_2 - 2}$$

$$= \frac{[\Sigma(x_1^2) - \bar{x}_1 \Sigma x_1] + [\Sigma(x_2^2) - \bar{x}_2 \Sigma x_2]}{n_1 + n_2 - 2}$$

where the counts (x), arithmetic mean ($\bar{x}$), and number of sampling units (n) are $x_1, \bar{x}_1, n_1$ for the first sample, and $x_2, \bar{x}_2, n_2$ for the second sample. The difference in sample means is significant at the 5% level ($P = 0.05$) when the calculated value of t is greater than the tabulated value for $2Q = 0.05$ and $v = n_1 + n_2 - 2$ degrees of freedom (see Table 12 in Pearson & Hartley 1966). 1% ($P = 0.01$) and 0.1% ($P = 0.001$) levels of significance are given in the columns headed $2Q = 0.01$ and $2Q = 0.001$.

A significant difference between variances will also produce a significant value of t. Therefore the t-test is often preceded by the F-test (section 7.1.1, method 2). The equality of variances is a necessary part of the null hypothesis, namely that the two samples are drawn from the same *normally-distributed* population. Therefore H_0 is rejected when there is a significant difference in means or variances. The basic requirement of the t-test is that the underlying parent distribution is at least approximately normal with variance independent of the mean. Therefore counts from contagious distributions must be transformed before the t-test is applied.

Example 26A. Comparison of two small samples from contagious distributions (F-test and t-test)

The power law was successfully fitted to a series of samples from contagious distributions ($s^2 > \bar{x}$). The estimate of the parameter b was 1·57, and therefore the transformation was $y = x^{0.2}$ (see examples 17 and 21). This transformation is now applied to the counts of two samples from the series. The two random samples were taken from a Population of *Gammarus pulex* at different times of the year. The dispersion of the Population was contagious on both occasions ($s^2 > \bar{x}$).

The counts of the first sample and their transformed values are given below (from example 21):

x	4	5	8	14	14	15	15	19	28	36
y	1·32	1·38	1·52	1·70	1·70	1·72	1·72	1·80	1·95	2·05

The statistics of the first sample are (from examples 15 and 21):

$\bar{x}_1 = 15\cdot80$, $s_1^2 = 99\cdot07$, $n_1 = 10$ before transformation,
$\bar{y}_1 = 1\cdot6860$, variance$_1$ = 0·0526 after transformation.

The 5 counts of the second sample are:

x	2	4	5	7	12
y	1·15	1·32	1·38	1·48	1·64

The statistics of the second sample are:

$\bar{x}_2 = 6\cdot00$, $s_2^2 = 14\cdot50$, $n_2 = 5$ before transformation,
$\bar{y}_2 = 1\cdot3940$, variance$_2$ = 0·0333 after transformation.

The null hypothesis (H_0) is that the two random samples come from normal populations with the same means and variances ($\mu_1 = \mu_2$, $\sigma_1^2 = \sigma_2^2$). The equality of variances is first tested by the F-test.

$$F = \frac{\text{variance}_1}{\text{variance}_2} = \frac{0\cdot0526}{0\cdot0333} = 1\cdot5796$$

Degrees of freedom, $v_1 = n_1 - 1 = 9$, $v_2 = n_2 - 1 = 4$.

The tabulated value of F is 8·90 for $v_1 = 9$ and $v_2 = 4$ (Pearson & Hartley 1966, Table 18, upper 2·5% points). As the calculated value of F is well below the tabulated value, the variances are not significantly different at the 5% level ($P > 0\cdot05$). Therefore H_0 is not disproved and the equality of means is now tested by the t-test. The variance of the total data is given by:

$$s^2 = \frac{[\Sigma(y_1^2) - \bar{y}_1 \Sigma y_1] + [\Sigma(y_2^2) - \bar{y}_2 \Sigma y_2]}{n_1 + n_2 - 2}$$

where $\bar{y}_1 = 1{\cdot}6860$, $\bar{y}_2 = 1{\cdot}3940$, $n_1 = 10$, $n_2 = 5$, y_1 are the 10 counts of sample 1, and y_2 are the 5 counts of sample 2. The standard deviation of the total data is given by

$$s = \sqrt{s^2} = 0{\cdot}215932 \quad \text{(the lengthy calculations are omitted).}$$

Therefore the standard error of difference of means

$$= s\sqrt{\frac{1}{n_1} + \frac{1}{n_2}} = 0{\cdot}215932 \sqrt{\frac{1}{10} + \frac{1}{5}} = 0{\cdot}11827$$

$$t = \frac{\text{difference between means}}{\text{standard error of difference}} = \frac{\bar{y}_1 - \bar{y}_2}{s\sqrt{\frac{1}{n_1} + \frac{1}{n_2}}}$$

$$= \frac{1{\cdot}6860 - 1{\cdot}3940}{0{\cdot}11827} = 2{\cdot}47$$

Degrees of freedom,

$$v = n_1 + n_2 - 2 = 13.$$

The tabulated value of t is $2{\cdot}16$ for $v = 13$ and $2Q = 0{\cdot}05$ (Pearson & Hartley 1966, Table 12). As the calculated value of t is greater than the tabulated value, the difference in sample means is significant at the 5% level ($P < 0{\cdot}05$). Note that the calculated value of t is less than the tabulated value of $3{\cdot}01$ for $2Q = 0{\cdot}01$, and therefore the means are not significantly different at the 1% level ($P > 0{\cdot}01$). Therefore H_0 is rejected at the 5% level but not at the 1% level. The difference in sample means is significant, but cannot be regarded as highly significant.

Example 26B. Comparison of two small samples from contagious distributions (one-way analysis of variance)

The F-test in the analysis of variance can be used instead of the t-test. A one-way analysis of variance is applied to the transformed counts of the two samples (see section 7.1.1, method 4 for all formulae). The transformed counts are given in Example 26A and the results are summarised in the usual analysis of variance table thus:

Source of variation	Sums of squares	Degrees of freedom	Mean square (variance)
Between samples	$S_2 = 0{\cdot}2841$	$i - 1 = 1$	$s_2^2 = 0{\cdot}2841$
Between sampling units within the samples (residual variance)	$S_1 = 0{\cdot}6063$	$N - i = 13$	$s_1^2 = 0{\cdot}0466$
Total	$S_T = 0{\cdot}8904$	$N - 1 = 14$	

The null hypothesis (H_0) is that both random samples come from normally-distributed populations with the same means ($\mu_1 = \mu_2$) and variances ($\sigma_1^2 = \sigma_2^2$).

Therefore s_1^2 and s_2^2 are estimates of the same variance. This hypothesis is tested by the F-test.

$$F = \frac{s_2^2}{s_1^2} = \frac{0.2841}{0.0466} = 6.10$$

with $v_1 = 1$ and $v_2 = 13$ degrees of freedom. The tabulated upper 5% point ($P = 0.05$) of F is 4.67 for $v_1 = 1$ and $v_2 = 13$ (Pearson & Hartley 1966, Table 18). As the calculated value of F is greater than the tabulated value, H_0 is rejected and the difference in sample means is significant at the 5% level ($P < 0.05$). Note that the calculated value of F is less than the tabulated upper 1% point ($P = 0.01$) of 9.07, and therefore the means are not significantly different at the 1% level ($P > 0.01$).

Compare the calculated values of F and t for the two samples and note that

$$\sqrt{F} = \sqrt{6.10} = 2.47 = t$$

Therefore, when there are only two samples in the analysis of variance, the relation $F = t^2$ holds (or $t = \sqrt{F}$), and the F-test is equivalent to the t-test which we used in Example 26A to compare the two means.

Therefore there is a choice between the t-test (preceded by the F-test on sample variances) and the F-test in the analysis of variance.

Example 27A. **Comparison of more than two samples from contagious distributions (one-way analysis of variance)**

The theory of this test and all definitions are given in section 7.1.1, method 4.

Bottom samples were taken at four stations in a stony stream and a random sample of 5 sampling units was taken at each station. Nymphs of the mayfly *Baëtis rhodani* were counted in each sampling unit and the following values were obtained:

Sample 1 (from station 1): 98, 22, 72, 214, 67; $\bar{x} = 94.60$, $s^2 = 5202.80$
Sample 2 (from station 2): 12, 13, 46, 38, 49; $\bar{x} = 31.60$, $s^2 = 320.30$
Sample 3 (from station 3): 86, 12, 49, 33, 72; $\bar{x} = 50.40$, $s^2 = 878.30$
Sample 4 (from station 4): 2, 5, 12, 3, 19; $\bar{x} = 8.20$, $s^2 = 51.70$

The variance was significantly greater than the mean ($s^2 > \bar{x}$) in each sample (see example 8B for χ^2 test on counts of sample 1). Therefore the nymphs were definitely clumped on the bottom of the stream, i.e. the dispersion of the Population was contagious at each station. As the conditions for the application of the analysis of variance are clearly not fulfilled, the counts were transformed to logarithms and the following values were obtained:

Sample	Count number					Total	Mean	Variance
	1	2	3	4	5			
1	1.991	1.342	1.857	2.330	1.826	9.346	1.8692	0.1268
2	1.079	1.114	1.663	1.580	1.690	7.126	1.4252	0.0918
3	1.935	1.079	1.690	1.519	1.857	8.080	1.6160	0.1158
4	0.301	0.699	1.079	0.477	1.279	3.835	0.7670	0.1663

$i = 4$, $n = 5$, $N = 20$, $\Sigma x = 28.387$, overall mean $\bar{x} = 1.4194$.

7.1 PARAMETRIC TESTS 109

The formulae for different sums of squares, degrees of freedom, and mean square (variance) are given in section 7.1.1, method 4. The results are summarised in the usual analysis of variance table thus:

Source of variation	Sums of squares	Degrees of freedom	Mean square
Between samples	$S_2 = 3\cdot333155$	$i - 1 = 3$	$s_2^2 = 1\cdot111051$
Between sampling units within the samples (residual variance)	$S_1 = 2\cdot002922$	$N - i = 16$	$s_1^2 = 0\cdot125182$
Total	$S_T = 5\cdot336077$	$N - 1 = 19$	

The null hypothesis (H_0) is that the four random samples come from normally-distributed populations with the same means ($\mu_1 = \mu_2 = \mu_3 = \mu_4$) and variances ($\sigma_1^2 = \sigma_2^2 = \sigma_3^2 = \sigma_4^2$). Therefore s_1^2 and s_2^2 are estimates of the same variance. This hypothesis is tested by the F-test.

$$F = \frac{s_2^2}{s_1^2} = \frac{1\cdot111051}{0\cdot125182} = 8\cdot875485$$

with $v_1 = 3$ and $v_2 = 16$ degrees of freedom. The tabulated upper 5% point ($P = 0\cdot05$) of F is 3·24 for $v_1 = 3$ and $v_2 = 16$ (Pearson & Hartley 1966, Table 18). As the calculated value of F is well above the tabulated value, H_0 is rejected at the 5% level, and also at the 1% level (tabulated value = 5·29), but *not* at the 0·1% level (tabulated value = 9·00). Therefore there are certainly significant differences between the means of the four samples ($P < 0\cdot01$ but $P > 0\cdot001$).

Example 27B. Checking the adequacy of the transformation

Before transformation, the variance increased with the mean for the four samples. Therefore we should check that the transformation has successfully established the independence of means and variances (see also section 3.2.4). A simple check is to plot variance against mean before and after transformation of the counts (see also example 7b and Fig. 6). Although there are only four points in the present example, this graphical method reveals a clear relationship between variance and mean before transformation but not after transformation (Fig. 20).

There are also several tests for homogeneity of variance and these can be applied to the sample variances. Bartlett's test is designed to test the null hypothesis that all the variance estimates being compared are estimates of the same variance (remember that one assumption for the analysis of variance is that the parent populations of the samples have the same variance, and therefore the separate within sample variances are estimates of the same variance). If there are i variance estimates s_i^2, each with f degrees of freedom ($f = n-1$), the test criterion is:

$$M = 2\cdot3026 f(i \log \bar{s}^2 - \Sigma \log s_i^2) \text{ where } \bar{s}^2 = \frac{\Sigma s_i^2}{i}$$

When some or all of the s_i^2 are less than 1, as in this example, the use of negative logarithms is avoided by multiplying all s_i^2 and $\bar{s}^2$ by 100 (or 1,000 if necessary).

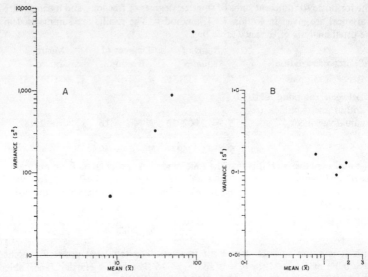

FIGURE 20. Relationship between mean and variance for samples in example 27, (A) before transformation, variance and mean tend to increase together, (B) after a logarithmic transformation, showing the relative independence of the variance from the mean. Values are plotted on log/log scale.

For the sample variances after transformation of counts:
$$M = 2 \cdot 3026 \, (4) \, (4 \log 12 \cdot 52 - 4 \cdot 3505) = 0 \cdot 3675$$
On the null hypothesis that each s_i^2 is an estimate of the same σ^2, M is distributed approximately as χ^2 with $i-1$ degrees of freedom. With small samples, Bartlett showed that M/C is more closely approximated by χ^2 with $i-1$ degrees of freedom, where
$$C = 1 + \frac{i+1}{3if} = 1 + \frac{4+1}{3(4)4} = 1 \cdot 1042$$
Therefore in the present example,
$$\frac{M}{C} = \frac{0 \cdot 3675}{1 \cdot 1042} = 0 \cdot 33$$
This value is clearly well below the tabulated χ^2 value of $7 \cdot 82$ ($P = 0 \cdot 05$) for $\nu = i-1 = 3$ degrees of freedom (Pearson & Hartley 1966, Table 8). Therefore the null hypothesis is accepted ($P > 0 \cdot 05$). As C is usually close to 1, it need be used only if M lies close to the 5% point of χ^2. When the degrees of freedom differ (i.e. value of n is not the same for each sample), different formulae must be used to calculate M and C (see Pearson & Hartley 1966, section 16).

As the calculation of M is comparatively laborious, the following quick, though less sensitive, tests can be used. We calculate either the ratio of largest to smallest variance estimates, or the ratio of largest to the sum of the variance estimates:

7.1 PARAMETRIC TESTS

Tables 31 and 31a in Pearson & Hartley (1966) give 5% and 1% points of these ratios for k independent variance estimates ($k = i$), each with v degrees of freedom ($v = f = n-1$). For the sample variances after transformation of counts:

$$\frac{s^2 \max}{s^2 \min} = \frac{0.1663}{0.0918} = 1.81.$$

This ratio clearly does not exceed the 5% value of 20·6 tabled for $k = 4$, $v = 4$. This result confirms the result of the M-test.

In applying these tests and the M-test, it must be remembered that the tests are also sensitive to departures from normality. Therefore a significant value may be due to non-normality rather than heterogeneity of variance. Whichever explanation is applicable, it can still be concluded that the transformation is inadequate when the above tests give a significant result.

Example 28. **Comparison of more than two samples from contagious distributions (two-way analysis of variance)**

The theory of this test and all definitions are given in section 7.1.1, method 4. The data are from example 27. The 20 sampling units were taken by 5 operators, and each operator took one sampling unit from each site. Operator 1 took all the sampling units in the column headed "Count number 1", operator 2 took all the units in the next column, and so on (see table of log counts in example 27). Therefore each count can be classified by sample number or operator number; e.g. 1·663 is the log of the third count in sample 2, and also the log of the second count taken by operator 3. The group totals are first found (one group for each operator):

	Count number				
Operator	1	2	3	4	5
Group total	5·306	4·234	6·289	5·906	6·652

All the basic information is now available for the calculation of sums of squares, etc. (see method 4 in section 7.1.1), and the results are summarised in the usual analysis of variance table:

Source of variation	Sum of squares	Degrees of freedom	Mean square
Between groups (operators)	$S_3 = 0.899375$	$n - 1 = 4$	$s_3^2 = 0.224844$
Between samples	$S_2 = 3.333155$	$i - 1 = 3$	$s_2^2 = 1.111051$
Residuals (error)	$S_1 = 1.103547$	12	$s_1^2 = 0.091962$
Total	$S_T = 5.336077$	$N - 1 = 19$	

For a two-way analysis of variance table with r rows and c columns, the mean squares estimate the following quantities:

Between columns	s_3^2	estimates	$\sigma_0^2 + r\sigma_c^2$
Between rows	s_2^2	estimates	$\sigma_0^2 + c\sigma_r^2$
Residuals	s_1^2	estimates	σ_0^2

where σ_c^2 is the component of variance due to variation between columns (i.e. between groups in this example), σ_r^2 is the component of variance due to variation between rows (i.e. between samples variation), and σ_0^2 is the error variance.

The null hypothesis (H_0) is that there are no real differences in column means (i.e. parent populations for groups have the same means and variances), or in row means (i.e. parent populations for samples have the same means and variances). The second part of H_0 is the same as H_0 in a one-way analysis of variance (see example 27A). If H_0 is correct, then each of the mean squares (s_3^2, s_2^2, s_1^2) is an independent estimate of the error variance σ_0^2.

Under H_0, the between groups component of variance is zero ($\sigma_c^2 = 0$). Therefore s_3^2 and s_1^2 are estimates of the same variance σ_0^2. This hypothesis is tested by the F-test.

$$F = \frac{s_3^2}{s_1^2} = \frac{0.224844}{0.091962} = 2.45$$

with $v_1 = 4$ and $v_2 = 12$ degrees of freedom. The tabulated upper 5% point ($P = 0.05$) of F is 3.26 for $v_1 = 4$ and $v_2 = 12$ (Pearson & Hartley 1966, Table 18). As the calculated value of F is well below the tabulated value, the mean squares are not significantly different at the 5% level ($P > 0.05$). Therefore there was no significant differences between the catches of the 5 operators.

Under H_0, the between samples component of variance is zero ($\sigma_r^2 = 0$). Therefore s_2^2 and s_1^2 are estimates of the same variance σ_0^2. This hypothesis is also tested by the F-test.

$$F = \frac{s_2^2}{s_1^2} = \frac{1.111051}{0.091962} = 12.08$$

with $v_1 = 3$ and $v_2 = 12$ degrees of freedom. This value is well above the upper 0.1% point of 10.80. Therefore there is a highly significant difference between the means of the four samples ($P < 0.001$).

It is thus possible to test independently for differences between samples (*e.g.* stations) and for differences between groups (*e.g.* operators). Although each group represents the catches of a different operator in this example, the different groups could also represent five types of sampler or five bottom types of each station.

7.2 NON-PARAMETRIC METHODS

In a parametric test (*e.g. t*-test), the null hypothesis (H_0) assumes a particular parent distribution (*e.g.* normal) and that parameters of this distribution (*e.g.* mean μ and variance σ^2) are the same for each sample. The following non-parametric methods do not require these assumptions, and can be applied to samples when the conditions for "normal" methods are not fulfilled. Therefore, non-parametric tests are an alternative to the use of "normal" methods on transformed counts (see section 3.2.4), and are particularly suitable for small samples from contagious distributions (cf. section 7.1.3). There is also the practical advantage that the calculations in non-parametric tests are usually simple, whereas the calculations in "normal tests" may be laborious, especially when data are transformed (see examples 26–28). Non-parametric tests are sometimes criticised because they do not utilise all the information provided by the sample. When the assumptions of a parametric test are valid, this test is more powerful than any other in rejecting H_0 when H_0 is false. Many non-parametric tests are almost as efficient as their parametric equivalents when all the conditions for the parametric test are fulfilled. When these conditions are not fulfilled, the non-parametric test is usually more powerful than its parametric equivalent.

The following tests were selected for their high power-efficiency and their suitability for small samples.

(1) *Comparison of two samples: the Mann–Whitney U-test*

This test is the non-parametric alternative to the *t*-test (section 7.1.3). The power-efficiency of the *U*-test is never less than 86%, is between 90% and 96% for normal data, and may be much higher than the efficiency of "normal" methods with non-normal data (Siegel 1956, Wilcoxon & Wilcox 1964).

The null hypothesis (H_0) is that two independent random samples are drawn from populations having the same parent distribution and the same means. Note that H_0 does not specify the form of the parent distribution, but simply assumes that it is the same for both samples. H_0 is tested by the Mann–Whitney *U*-test, which is a test of rank order, *i.e.* the counts are replaced by their rank values in a single sequence.

The procedure is as follows:

1. Consider counts in both samples together and arrange these counts in a single array from lowest to highest. n_1 = number of counts in sample 1, n_2 = number of counts in sample 2.
2. Substitute a rank for each count. Ranks range from 1 for the lowest count to N for the highest count ($N = n_1 + n_2$). If any counts are equal, they are given the average of the tied ranks.
3. Total the ranks for each sample.
   R_1 = sum of ranks in sample 1, R_2 = sum of ranks in sample 2.
   Check that $R_1 + R_2 = \dfrac{(n_1+n_2)(n_1+n_2+1)}{2}$
4. Calculate test statistics U_1 and U_2:
$$U_1 = n_1 n_2 + \frac{n_2(n_2+1)}{2} - R_2$$
$$U_2 = n_1 n_2 + \frac{n_1(n_1+1)}{2} - R_1$$

5. Refer the smaller of the two values, U_1 and U_2, to the appropriate value of U in Table 14 (located by values of n_1 and n_2). If the calculated value of U is equal to or less than the tabulated value, H_0 is rejected at the 5% level of significance ($P = 0.05$). Note that *small* values of U cause rejection.

TABLE 14. VALUES OF THE U-STATISTIC AT THE 5% LEVEL OF SIGNIFICANCE.

n_1 and n_2 are the number of counts in each sample. Remember that small values of U cause rejection of H_0 at the 5% level.

n_1 \ n_2	2	3	4	5	6	7	8	9	10	11	12	13	14	15	16	17	18	19	20
2							0	0	0	0	1	1	1	1	1	2	2	2	2
3				0	1	1	2	2	3	3	4	4	5	5	6	6	7	7	8
4			0	1	2	3	4	4	5	6	7	8	9	10	11	11	12	13	13
5		0	1	2	3	5	6	7	8	9	11	12	13	14	15	17	18	19	20
6		1	2	3	5	6	8	10	11	13	14	16	17	19	21	22	24	25	27
7		1	3	5	6	8	10	12	14	16	18	20	22	24	26	28	30	32	34
8	0	2	4	6	8	10	13	15	17	19	22	24	26	29	31	34	36	38	41
9	0	2	4	7	10	12	15	17	20	23	26	28	31	34	37	39	42	45	48
10	0	3	5	8	11	14	17	20	23	26	29	33	36	39	42	45	48	52	55
11	0	3	6	9	13	16	19	23	26	30	33	37	40	44	47	51	55	58	62
12	1	4	7	11	14	18	22	26	29	33	37	41	45	49	53	57	61	65	69
13	1	4	8	12	16	20	24	28	33	37	41	45	50	54	59	63	67	72	76
14	1	5	9	13	17	22	26	31	36	40	45	50	55	59	64	67	74	78	83
15	1	5	10	14	19	24	29	34	39	44	49	54	59	64	70	75	80	85	90
16	1	6	11	15	21	26	31	37	42	47	53	59	64	70	75	81	86	92	98
17	2	6	11	17	22	28	34	39	45	51	57	63	67	75	81	87	93	99	105
18	2	7	12	18	24	30	36	42	48	55	61	67	74	80	86	93	99	106	112
19	2	7	13	19	25	32	38	45	52	58	65	72	78	85	92	99	106	113	119
20	2	8	13	20	27	34	41	48	55	62	69	76	83	90	98	105	112	119	127

6. Table 14 cannot be used when n_1 or n_2 is greater than 20. As n_1 and n_2 increase in size, the sampling distribution of U rapidly approaches the normal frequency distribution, with mean $= \dfrac{n_1 n_2}{2}$ Therefore calculate normal deviate d:

$$d = \frac{U - (n_1 n_2)/2}{\sqrt{\dfrac{n_1 n_2 (n_1 + n_2 + 1)}{12}}}$$

H_0 is rejected at 5% level ($P = 0.05$) when absolute value of d is greater than 1·96, at 1% level ($P = 0.01$) when $d > 2.58$, and at 0·1% level ($P = 0.001$) when $d > 3.29$.

Example 29. Comparison of two small samples from contagious distributions (Mann–Whitney U-test)

The following counts of *Gammarus pulex* were obtained from two small random samples (counts are arranged from lowest to highest in each sample).

Sample 1: 2, 4, 5, 7, 12; $n_1 = 5$
Sample 2: 4, 5, 8, 14, 14, 15, 15, 19, 28, 36; $n_2 = 10$

These samples were compared in example 26, but the counts had to be transformed before the *t*-test was applied. The samples are now compared by the Mann–Whitney *U*-test. Each count is given its rank in the joint array thus:

Sample 1: 1, 2½, 4½, 6, 8
Sample 2: 2½, 4½, 7, 9½, 9½, 11½, 11½, 13, 14, 15

Note that ranks range from 1 for lowest count of 2, to 15 for highest count of 36 ($N = 15 = n_1 + n_2$). All equal counts are given the same rank, which is the average of the appropriate ranks in the array; *e.g.* the two counts 4 lie over ranks 2 and 3, and therefore each count is ranked 2½.

Sum of ranks in sample 1 = $R_1 = 22$
Sum of ranks in sample 2 = $R_2 = 98$

Check $R_1 + R_2 = 120 = \dfrac{(n_1+n_2)(n_1+n_2+1)}{2}$

$$U_1 = n_1 n_2 + \frac{n_2(n_2+1)}{2} - R_2$$

$$= 50 + \frac{110}{2} - 98 = 7$$

$$U_2 = n_1 n_2 + \frac{n_1(n_1+1)}{2} - R_1$$

$$= 50 + \frac{30}{2} - 22 = 43$$

Check that: $U_1 + U_2 = 50 = n_1 n_2$. Therefore smallest calculated value of U is 7. Tabulated value of U is 8 for $n_1 = 5$ and $n_2 = 10$ (see Table 14). As calculated value of U is less than tabulated value, H_0 is rejected at the 5% level of significance ($P < 0.05$). Therefore the mean level of sample 2 is significantly higher than that of sample 1 ($P < 0.05$). The calculated value of U is just below the tabulated value at the 5% level, and reference to more detailed tables (*e.g.* Siegel 1956, Table K) reveals that U is not significant at the 2% level ($P > 0.02$). Therefore the *U*-test is just as sensitive as the *t*-test, which indicated a significant difference in means at the 5% level but not at the 1% level (see example 26A).

(2) *Comparison of more than two samples* (*one-way classification*)

These tests are the non-parametric alternatives to the one-way analysis of variance (see section 7.1.1, method 4, and examples 26B, 27A).

(A) *Quick method* (Quenouille 1959). The power-efficiency of this test is about 81% for normal distributions, but may be higher for some non-normal distributions. The null hypothesis (H_0) is that all samples come from the same population, and therefore there is no difference in mean level between several samples. Note that this test is independent of every hypothesis about the population variances. Each sample should have at least 5 sampling units (*i.e.* 5 counts). The procedure is as follows:

1. Consider counts of all samples together and estimate:

    $y(\frac{1}{4})$ = count exceeded by $\frac{1}{4}$ of total counts
    $y(\frac{3}{4})$ = count exceeded by $\frac{3}{4}$ of total counts

2. In each sample, find the number of counts that exceed $y(\frac{1}{4})$ or fall below $y(\frac{3}{4})$. Call these numbers a_i and b_i for each sample.

3. Calculate χ^2 thus:

$$\chi^2 = \sum_i \frac{(a_i - b_i)^2}{a_i + b_i}$$

4. Refer χ^2 to Table 8 in Pearson & Hartley (1966). H_0 is accepted if calculated χ^2 value is less than tabulated value in column headed $Q = 0.050$ (for 5% significance level), and $v = i-1$ degrees of freedom (i = number of samples). H_0 is rejected at 5% level when calculated χ^2 greater than tabulated value in column headed $Q = 0.050$, and at 1% level when calculated $\chi^2 >$ tabulated value in column headed $Q = 0.010$ (cf. χ^2 test for "goodness-of-fit", section 4.1.3).

Example 30. **Quenouille's test of the differences in mean level between several samples**

Four random samples of nymphs of *Baëtis rhodani* were compared in example 27, but the counts had to be transformed before a one-way analysis of variance was applied. The counts are given below and are arranged from lowest to highest in each sample (i = number of samples = 4):

Sample 1: 22, 67, 72, 98, 214
Sample 2: 12, 13, 38, 46, 49
Sample 3: 12, 33, 49, 72, 86
Sample 4: 2, 3, 5, 12, 19

Count exceeded by $\frac{1}{4}$ of total counts = $y(\frac{1}{4})$ = 71
Count exceeded by $\frac{3}{4}$ of total counts = $y(\frac{3}{4})$ = $12\frac{1}{2}$

Number of counts above $y(\frac{1}{4})$ ($=a_i$) and below $y(\frac{3}{4})$ ($=b_i$) in each sample are given below together with χ^2:

Sample	a_i	b_i	χ^2
1	3	0	9/3 = 3·00
2	0	1	1/1 = 1·00
3	2	1	1/3 = 0·33
4	0	4	16/4 = 4·00

Total $\chi^2 = 8.33$ with $v = i - 1 = 3$ degrees of freedom.

As this χ^2 value is just above the 5% point of 7·81 ($Q = 0·050$ and $v = 3$) in Pearson & Hartley (1966) Table 8, H_0 is rejected at the 5% level ($P < 0·05$), but not at the 1% level (tabulated value = 11·34 and therefore $P > 0·01$).

(B) *Kruskal–Wallis one-way analysis by ranks.* The power-efficiency of this test is about 96% (Siegel 1956). The null hypothesis (H_0) is that all samples come from the same population, and therefore there is no difference in mean level between several samples. The number of counts in each sample can be different. The procedure is as follows:

1. Consider counts of all samples together and arrange these counts in a single array from lowest to highest.

2. Substitute a rank for each count. Ranks range from 1 for lowest count to N for the highest count (N = total number of counts = Σn_i where n_i = number of counts in each sample). Equal counts are given the average of the tied ranks.

3. Total the ranks for each sample (i = number of samples).

$$R_1 = \text{sum of ranks in sample 1}$$
$$R_2 = \text{sum of ranks in sample 2}$$
$$R_i = \text{sum of ranks in sample } i$$

4. Calculate test statistic K:

$$K = \frac{12}{N(N+1)} \sum \frac{(R_i^2)}{n_i} - 3(N+1)$$

where

$$\sum \frac{(R_i^2)}{n_i} = \frac{R_1^2}{n_1} + \frac{R_2^2}{n_2} + \cdots + \frac{R_i^2}{n_i}$$

K is distributed approximately as χ^2 with $v = i-1$ degrees of freedom.

5. Refer K to Table 8 in Pearson & Hartley (1966). H_0 is rejected at 5% level when K greater than tabulated value of χ^2 in column headed $Q = 0·050$ (see Quenouille's test and example 30). When there are only three samples ($i = 3$) and the number of counts in each of the three samples is 5 or fewer, the χ^2 approximation cannot be used and K must be referred to special tables (see Siegel 1956, Table O; Campbell 1967, Table A4).

Example 31. Kruskal–Wallis test for difference in mean level between several samples

The four samples of example 30 (see also example 27) are also used in this example. Each count is replaced by its rank in the whole array thus:

Sample	Ranks					n_i	R_i	$\dfrac{R_i^2}{n_i}$
1	9	15	16½	19	20	5	79½	1264·05
2	5	7	11	12	13½	5	48½	470·45
3	5	10	13½	16½	18	5	63	793·80
4	1	2	3	5	8	5	19	72·20

$$N = 20 \quad R = 210 \quad \sum \frac{(R_i^2)}{n_i} = 2600·50$$

$$K = \frac{12}{20(21)}(2600·5) - 3(21)$$
$$= 74·3 - 63$$
$$= 11·3$$

This value of K is referred to tables of the χ^2 distribution for $v = i-1 = 3$ degrees of freedom. As K is well above the 5% point of 7·81 ($Q = 0·050$ and $v = 3$) in Pearson & Hartley (1966) Table 8, H_0 is rejected at the 5% level ($P < 0·05$). H_0 is just rejected at the 1% level (tabulated value = 11·3 and therefore $P = 0·01$).

As example 30 and 31 use the same samples, they illustrate the relative power-efficiency of the two tests. The Kruskal–Wallis test rejects H_0 just on the 1% level ($P = 0·01$), whereas Quenouille's test rejects H_0 only at the 5% level (P between 0·05 and 0·01). Therefore there is a greater chance of making a Type II error (*i.e.* failure to reject H_0 when H_0 is false), when Quenouille's test is used.

(3) *Comparison of more than two samples (two-way classification): Friedman two-way analysis by ranks.* Little is known about the power of Friedman's test, but the power-efficiency is probably close to that of the parametric F-test (Siegel 1956). The number of counts in each sample must be the same. Each count belongs to one sample and also one group, where the groups could represent different bottom types, different samplers, different workers, etc. (see example 28 and page 101). The null hypothesis (H_0) is that there is no difference in mean level between groups. The procedure is as follows:

1. Arrange counts in a two-way table with i samples (rows) and n groups (columns).
2. Rank counts in each row from 1 to n.
3. Determine sum of ranks in first column = R_1, in second column = R_2, in nth column = R_n.
4. Determine total sum of ranks, $\Sigma R_n = R_1 + R_2 + \cdots + R_n$.

5. Calculate S thus:
$$S = \Sigma(R_n^2) - \frac{(\Sigma R_n)^2}{n}$$
where $\Sigma(R_n^2) = R_1^2 + R_2^2 + \cdots + R_n^2$

6. Refer S to the appropriate value of S in Table 15 (located by values of n and i). If calculated value of S is greater than or equal to tabulated value, then H_0 is rejected at the stated level of significance (approximate 5%, 1%, 0·1% points are given in Table 15).

TABLE 15. VALUES OF THE S-STATISTIC AT THE 5%, 1%, AND 0·1% LEVELS OF SIGNIFICANCE.

i = number of samples, n = number of groups. Significance levels are only approximate.

n	3			4			5		
Significance level	5%	1%	0·1%	5%	1%	0·1%	5%	1%	0·1%
i									
3	18			37			64	76	86
4	26	32		52	64	74			
5	32	42	50	65	83	105			
6	42	54	72	76	100	128			
7	50	62	86						
8	50	72	98						
9	56	78	114						
10	62	96	126						

7. For values of n or i greater than those given in Table 15, calculate χ^2 thus:
$$\chi^2 = \frac{12S}{in(n+1)}$$
Refer this χ^2 value to tables of the χ^2 distribution with $v = n - 1$ degrees of freedom (Pearson & Hartley 1966, Table 8). H_0 is rejected at 5% level when calculated χ^2 greater than tabulated value in column headed $Q = 0.050$ (see examples 30, 31).

When the Kruskal–Wallis and Friedman tests are combined, comparisons can be made between samples and between groups for the same set of counts. Although these analyses by ranks are an alternative to the two-way analysis of variance, they provide less information than the parametric test. The non-parametric analyses only indicate differences between samples or between groups, whereas the parametric analysis partitions the total variation in a set of counts and can be extended to tests for interactions between different sources of variation.

Example 32 **Friedman's test for differences in mean level between several samples**
The four samples of examples 27, 28, 30, 31 are also used in this example. The counts are classified by sample number and by group number. A different operator took the 4 counts in each group (see example 28).

	\multicolumn{5}{c}{Group number}
	1	2	3	4	5
Sample 1	98	22	72	214	67
Sample 2	12	13	46	38	49
Sample 3	86	12	49	33	72
Sample 4	2	5	12	3	19

There are four samples ($i = 4$), and five groups ($n = 5$). H_0 is that there is no difference in mean level between groups, i.e. the use of different operators did not affect the size of the catch. Each count is now replaced by its rank in each row thus:

	\multicolumn{5}{c}{Group number}	
Sample 1	4	1	3	5	2	
Sample 2	1	2	4	3	5	
Sample 3	5	1	3	2	4	
Sample 4	1	3	4	2	5	
Totals	11	7	14	12	16	$\Sigma R_n = 60$
	R_1	R_2	R_3	R_4	R_5	$\Sigma(R_n^2) = 766$

Check that $\Sigma R_n = 60 = \dfrac{in(n+1)}{2}$

$$S = \Sigma(R_n^2) - \frac{(\Sigma R_n)^2}{n} = 766 - \frac{60^2}{5} = 46$$

As $n = 5$ and $i = 4$, Table 15 cannot be used. Therefore:

$$\chi^2 = \frac{12S}{in(n+1)} = \frac{12(46)}{120} = 4 \cdot 6$$

This χ^2 value is referred to tables of the χ^2 distribution with $v = n-1 = 4$ degrees of freedom. As the calculated χ^2 value is well below the 5% point of 9·49 ($Q = 0.050$ and $v = 4$) in Pearson & Hartley (1966) Table 8, H_0 is not rejected at the 5% level ($P > 0.05$). Therefore there was no significant difference between the catches of the 5 operators. A similar conclusion was reached in example 28.

(4) *Correlation between two variables.* Use of the parametric correlation coefficient (r) requires a bivariate normal distribution. Therefore counts from bottom samples are usually transformed before r is calculated (section 7.1.1, method 5). The non-parametric alternative is a rank correlation coefficient (power-efficiency = 91%). If there are n pairs of observations with values x and y in each pair, then the first step is to rank x and y separately. Next calculate the difference

(d) between each pair of ranked values (check $\Sigma d = 0$) and square each difference (d^2). Calculate Spearman's rank correlation coefficient (r_s):

$$r_s = 1 - \frac{6\Sigma d^2}{n(n^2-1)}$$

r_s ranges from -1 (complete discordance between ranked values) to $+1$ (complete concordance). The correlation is significant at the 5% level ($P = 0.05$) when r_s equals or exceeds 1.00 for $n = 5$, 0.89 for $n = 6$, 0.75 for $n = 7$, 0.71 for $n = 8$, 0.68 for $n = 9$, 0.65 for $n = 10$. For $n > 10$, the null distribution of r_s is similar to that of r, and the correlation is significant at the 5% level when r_s exceeds the tabulated value of r for $v = n-2$ degrees of freedom and $2Q = 0.05$ (Pearson & Hartley 1966, Table 13). Note that r_s is never an estimator of r, even for large samples.

Other non-parametric correlation coefficients are Kendall's tau coefficient, Kendall's coefficient of partial rank correlation, and the coefficient of concordance (Kendall 1962, Pearson & Hartley 1966, section 25). As a full description of both parametric and non-parametric methods of correlation would require another booklet, only a brief outline of these methods is included in the present text.

(5) χ^2 *and contingency tables.* The χ^2 distribution has many uses in statistics. Its use to test for agreement between observation and hypothesis can be extended from goodness-of-fit tests (section 4.1.3) to any situation where a basic hypothesis specifies the proportions or probabilities of a series of observations falling into several groups. A simple example with only two groups is the distribution of sexes. For example, there are 53 males and 67 females in a total catch of 120 animals. The null hypothesis (H_0) assumes that there are equal numbers of each sex, *i.e.* expected values are 60 and 60. Therefore:

$$\chi^2 = \frac{(\text{Observed}-\text{expected})^2}{\text{expected}} = \frac{(53-60)^2}{60} + \frac{(67-60)^2}{60} = 1.63$$

with $v = k-1 = 2-1 = 1$ degree of freedom ($k =$ number of groups). As calculated χ^2 value is well below tabulated value of 3.84 ($v = 1$ and $Q = 0.050$) in Pearson & Hartley (1966) Table 8, H_0 is not rejected at the 5% level ($P > 0.05$). Therefore there is no evidence of a significant departure from the expected equality of the sexes.

This method can be extended to k groups (see example 33). When $k = 2$, each expected value should be at least 5. When $k > 2$, only 20% of the expected frequencies should be less than 5 and no expected frequencies should be less than 1 (Cochran 1954). As it is often difficult to estimate the expected values under H_0, this χ^2 test is of limited value. The usual H_0 is that all expected values are the same, but this assumption is rarely justified.

Example 33. χ^2 **test for a single classification** ($k > 2$)

We wish to compare the efficiency of 5 traps ($k = 5$), and the observed catches are 6, 8, 16, 5, 18. H_0 is that all traps are equally efficient and therefore the expected catches are the same

$$\text{expected} = \frac{\text{total catch}}{k} = \frac{53}{5} = 10.6.$$

Therefore:

$$\chi^2 = \frac{(6-10\cdot6)^2}{10\cdot6} + \cdots + \frac{(18-10\cdot6)^2}{10\cdot6} = \frac{143\cdot2}{10\cdot6} = 13\cdot51$$

with $v = k - 1 = 4$ degrees of freedom. As this χ^2 value is well above the tabulated value of 9·49 ($v = 4$ and $Q = 0.050$) in Pearson & Hartley (1966) Table 8, H_0 is rejected at the 5% level ($P < 0.05$) and also at the 1% level (tabulated value = 13·28, $P < 0.01$). Therefore the traps are not equally efficient.

When individuals are classified in two directions with two or more categories in each classification, the data are arranged in a *contingency table* with k columns and r rows. The individuals in each cell of the table belong to the same category in each system of classification. The null hypothesis (H_0) is that the column proportions are equal, or alternatively that the row proportions are equal. Expected values under H_0 are calculated for each cell of the table. Observed and expected values for each cell are compared by the usual χ^2 test, and the total χ^2 for the whole table is referred to appropriate tables of χ^2 with $v = (k-1)(r-1)$ degrees of freedom. As the χ^2 statistic is the same for a test of either null hypothesis, it is a good test of association between the two classifications. Therefore the χ^2 test is used to detect any difference between rows in proportions from column to column, or any difference between columns in the proportions from row to row.

The simplest table is used to compare the proportions in two independent samples, and has only two rows and columns (see example 34). A 2 by 2 table can also be used to test for association between two species (Greig–Smith 1964, Chapter 4). More complex tables have k columns and r rows (see example 35). When either k or r is larger than 2, fewer than 20% of the cells should have expected values of less than 5 and no cell should have an expected value of less than 1 (Cochran 1954). If these conditions are not fulfilled, the values in adjacent cells must be combined. Expected values should be at least 5 in a 2 by 2 contingency table. If this condition is not fulfilled,

7.2 NON-PARAMETRIC METHODS

it is obviously impossible to combine adjacent values and an exact test must be used (Bailey 1959, Chapter 7; Pearson & Hartley 1966, section 19). The theory of χ^2 assumes that the numbers involved in a contingency table are really continuously variable and not discrete. Therefore Yates' correction is used in a small contingency table (see example 34).

Example 34. Use of a 2 by 2 contingency table

Several bottom samples were taken near and away from the banks in a small stony stream. Equal numbers of sampling units were taken in each group of samples. Nymphs of a mayfly were counted in each group of samples, and the last instars were counted separately from the other instars. The counts are given in the following contingency table together with the usual symbols.

Samples	Last instar	Other instars	Totals
near bank	$a = 42$	$b = 96$	$a + b = 138$
away from bank	$c = 12$	$d = 104$	$c + d = 116$
Totals	$a + c = 54$	$b + d = 200$	$n = a + b + c + d$
			$= 254$

The null hypothesis (H_0) is that the proportions in the two groups of samples are the same, i.e. there is no association between the two classifications. χ^2 could be calculated by first finding expected values for the 4 cells, and then comparing observed and expected values in each cell. It is easier to use the following short-cut formula for a 2 by 2 table:

$$\chi^2 = \frac{n(|ad-bc|-n/2)^2}{(a+b)(c+d)(a+c)(b+d)}$$

where $|ad-bc|$ means the absolute value of $(ad-bc)$, and $n/2$ is Yates' correction for continuity.

$$\chi^2 = \frac{254(|(42)(104)-(96)(12)|-127)^2}{(138)(116)(54)(200)} = \frac{2,423,647,934}{172,886,400}$$

$$= 14 \cdot 02 \text{ with } v = (k-1)(r-1) = 1 \text{ degree of freedom.}$$

As this χ^2 value is above the 0·1% point of 10·83 in Pearson & Hartley (1966) Table 8, the association between the two classifications is highly significant ($P < 0.001$). Therefore H_0 is rejected at the 0·1% level, and there is strong evidence that the proportion of last instars is significantly higher near the bank.

Example 35. Use of a large contingency table

Bottom samples were taken at four sites in a small stream, and larvae of *Sericostoma personatum* (Trichoptera) were counted in each sample. There were six larval instars, and the total catches of each instar at each site were arranged in a contingency table (Table 16) with six columns ($k = 6$) and four rows ($r = 4$). The null hypothesis (H_0) is that the proportion of larvae occurring in each instar is the same at the four sites, i.e. there is no association between the two classifications. Expected values were calculated for each cell of the table. These expected catches would occur if H_0 was correct. To find the expected value for each cell, divide the product of the corresponding row and column totals by the grand total for the whole table, e.g. for second cell in second row, observed catch = 12 and

$$\text{expected catch} = \frac{\text{(row total) (column total)}}{\text{grand total}} = \frac{(148)(109)}{978} = 16.50.$$

Observed and expected values for each cell are compared by the usual χ^2 test and the total χ^2 for the whole table is:

$$\chi^2 = \sum \frac{(\text{Observed} - \text{expected})^2}{\text{expected}} = \frac{(26-7.68)^2}{7.68} + \cdots + \frac{(82-59.97)^2}{59.79}$$
$$= 205.09 \text{ with } v = (k-1)(r-1) = 15 \text{ degrees of freedom.}$$

TABLE 16. TOTAL CATCHES OF EACH INSTAR AT EACH SITE.

Numbers in parentheses are the expected values for χ^2 analysis.

Instar	1	2	3	4	5	6	Total
Site 1	26	58	56	44	35	40	259
	(7.68)	(28.87)	(45.02)	(36.81)	(49.26)	(91.37)	
Site 2	2	12	39	27	38	30	148
	(4.39)	(16.50)	(25.73)	(21.03)	(28.15)	(52.21)	
Site 3	1	30	54	56	67	193	401
	(11.89)	(44.69)	(69.70)	(56.99)	(76.26)	(141.46)	
Site 4	0	9	21	12	46	82	170
	(5.04)	(18.95)	(29.55)	(24.16)	(32.33)	(59.97)	
Total	29	109	170	139	186	345	978

As this χ^2 value is well above the 0.1% point of 37.70 in Pearson & Hartley (1966) Table 8, the association between the two classifications is highly significant ($P < 0.001$). Therefore H_0 is rejected at the 0.1% level. The χ^2 values for each cell indicate that instars 1 and 2 were more abundant than expected at site 1 and least abundant at sites 3 and 4, whereas the reverse was true for instars 5 and 6.

7.3 SUMMARY GUIDE

All the parametric and non-parametric tests described in section 7 are listed in Table 17. This table serves as a guide to the various methods and indicates the example in which each test is described.

TABLE 17. PARAMETRIC AND NON-PARAMETRIC TESTS FOR THE COMPARISON OF SAMPLES

	Parametric test	Transformation necessary?	Example	Non-parametric test	Example
Comparison of two samples—					
A. Means of large samples ($n > 50$)	Normal deviate (d)	No	23	U-test	29
Small samples ($n < 50$) from a:					
(1) random distribution	d for Poisson	No	25		
	t-test	Yes	26	U-test	29
(2) contagious distribution	F-test	No	24		
B. Variances of large samples ($n > 50$)	F-test	Yes	26	U-test	29
Small samples ($n < 50$)					
Comparison of 3+ samples from random or contagious distributions	Analysis of variance	Yes	27	Quenouille	30
			28	Kruskal–Wallis	31
				Friedman	32
Correlation between two variables (dependent variable y and independent variable x)	Correlation coefficient	Yes (both x and y)		Rank correlation coefficient	
	Regression analysis	Yes (only y)			
Correlation between 3+ variables:					
Variables analysed in pairs	Regression analysis	Yes (only y)			
Variables analysed simultaneously	Multiple regression	Yes			
"Goodness-of-fit" tests with specified hypothesis				χ^2 one-way	33
Association between two classifications:					
Comparison of proportions in two independent samples				2 by 2 contingency table	34
Comparison of proportions in several categories				k by r contingency table	35

VIII   PLANNING A SAMPLING PROGRAMME

The statistical principles of sampling are the subject of several books (*e.g.* Yates 1960, Stuart 1962, Cochran 1963), and only some aspects are discussed in this section. It is first necessary to define clearly the objects of the study and the area to be sampled. The frequency of sampling will depend upon the objects of the study. Samples may be taken at weekly intervals in detailed studies of life histories, or only once a year in some general surveys. Most investigations are either extensive faunal surveys or intensive quantitative studies. These two broad categories are now discussed separately.

8.1 Faunal surveys

A survey of a large sampling area (lake, river, etc.) usually precedes a quantitative investigation, but may be an end in itself. The whole sampling area is usually divided into sections of equal area, and one or more stations are selected in each section. Whenever possible, the stations should be selected at random. Alternatively the sampling area is divided into different biotopes (*e.g.* stony substratum, mud, moss, etc.) and stations are selected at random in each biotope.

The chief objects of a faunal survey are to discover which species are present, and to estimate the relative abundance of each species at each station. As it is important to catch the rare species at each station, each sample should cover a large area of bottom. A simple method is to collect over a large sampling unit with a pond net for a fixed period of time. This technique can be used in flowing water and still water. Collections in rivers are obtained by disturbing the bottom, either by hand-turning of stones or by kicking with the feet, and allowing the current to carry the dislodged material into a collecting net. The collector slowly moves upstream for 2 minutes and thus covers a large area of bottom. A similar technique can be used on the stony substratum of lake shores. Stones are lifted and cleaned in the mouth of the net which is quickly swept underneath the stones. The collector slowly moves along the lake shore for 2 minutes. More than one collection may be necessary when a station covers a large area of bottom.

The advantages of these simple techniques are: (1) they do not require elaborate apparatus; (2) they usually catch a high proportion of the total species present at each station; and (3) they often provide fairly comparable figures, especially when the biotope and collector are the same for all samples; it is also possible to compare the percentage composition of the benthos at each station.

The disadvantages are: (1) they cannot be used in deep water; (2) as collections are taken on a time basis, the area of the sampling unit is not fixed. Therefore the samples do not provide estimates of numbers per unit area, and comparisons between samples are limited to relative abundance. As the expression of the numbers of each species as a percentage of total numbers is strongly influenced by numbers in the rest of the samples, comparisons of percentage composition are of limited value.

If a quantitative technique is preferred to the above methods, there is a choice of numerous samplers (Welch 1948, Macan 1958, Southwood 1966). Although the size of the sampling unit is known with these techniques, it usually covers a small area of bottom. Therefore the major problem is to decide how many sampling units are necessary to ensure that the sample includes most of the species present. If the sampling units are taken in a long transect line which runs parallel to some obvious environmental gradient, there is a high probability that most species will be taken at least once. When species can be identified in the field, the sampling units are taken at random and the number of new species is noted in each successive sampling unit. No more sampling units are taken when three successive units have failed to add any species to the total list. The number of sampling units required at each station will depend upon the diversity and dispersion of the bottom fauna. Gaufin, Harris & Walter (1956) give a formula for calculating the average number of new species contributed by the Kth successive sample ($K = 1$ or more), and Harris (1957) proposes a formula for computing the standard errors of such estimates.

8.2 Quantitative studies

The methods described in sections 4–7 are all used in quantitative studies and require a knowledge of numbers per unit area. Some problems of quantitative sampling were mentioned in previous sections, and the major considerations are (1) the dimensions of the sampling unit (quadrat size), (2) the number of sampling units in each sample, and (3) the location of sampling units in the sampling area. It is often impossible to make a complete and accurate estimate of the numbers of all species in a large area of bottom. Therefore most quantitative investigations are restricted to a study of a small number of species in a large area, or a larger number of species in a small area.

It is important to define the area in which the investigation takes place; thus if only a section of a lake or river is sampled, then the area of this section must be clearly defined.

8.2.1 *The dimensions of the sampling unit (quadrat size)*

The defined area is divided up into sampling units of equal size, and the whole aggregate of sampling units forms the population (statistical meaning, see section 2.1). Therefore the total number of available sampling units in the population depends upon the relationship between the area of the population (= total sampling area) and the area of the sampling unit (= quadrat size).

A small quadrat size is most suitable for a study of the dispersion of a Population (ecological meaning, see section 2.1), and the problems of detecting non-randomness were fully discussed in section 5.4. If the dispersion of a Population is truly random, all quadrat sizes are equally efficient in the estimation of population parameters. Efficiency is defined in terms of the relative amounts of sampling required to give estimates of equal precision. Several workers (*e.g.* Beall 1939, Finney 1946, Taylor 1953) have investigated the effects of the size of the sampling unit on the efficiency of sampling, and they conclude that a small unit is more efficient than a larger one when the dispersion of a Population is contagious. The advantages of a small sampling unit over a larger unit are: (1) more small units can be taken for the same amount of labour in dealing with the catch; (2) as a sample of many small units has more degrees of freedom than a sample of a few large units, the statistical error is reduced; and (3) since many small units cover a wider range of the habitat than a few large units, the catch of the small units is more representative. In general, the smaller the sampling units employed, the more accurate and representative will be the results.

Although the ideal solution is to use the smallest possible sampling unit, many practical factors will set a lower limit to the dimensions of the sampling unit, *e.g.* stone size will be a limiting factor on a stony substratum. It must also be remembered that with a small sampling unit, the sampling error at the edge of the unit is proportionally greater. Therefore the choice of the final quadrat size is always a compromise between statistical and practical requirements.

8.2.2 *The number of sampling units in each sample*

As the dispersion of many species is frequently contagious, a large variation is encountered in sampling natural populations and small samples are statistically inaccurate. Therefore published quantitative data are often unreliable because the samples are too small. The simplest solution to this problem is to always take very large samples, *i.e.* with over 50 sampling units in each sample ($n > 50$). Unfortunately, it is usually impossible to sort and count all the different species in a very large sample, especially when samples are taken at frequent intervals. Therefore a compromise must be made between statistical accuracy and labour.

The following simple method can be used to find a suitable number of sampling units. Take 5 sampling units at random and calculate the arithmetic mean. Next take 5 more units at random and calculate the mean for 10 units. Continue to increase sample size by 5-unit steps, and plot means for 5, 10, 15, etc. units against sample size. When the mean value ceases to fluctuate, a suitable sample size has been reached and this sample size can be used for that particular station. As it is often impossible to calculate means at the time of sampling, this simple method is of limited application.

Sample size can be calculated for a specified degree of precision. First decide how large an error can be tolerated in the estimate of the population mean. The percentage error can be expressed as either the standard error of the mean (section 6.1), or confidence limits of the mean (section 6.2). For a given standard deviation (or variance s^2), the standard error is a function of the number (n) of sampling units in each random sample. The ratio of standard error to arithmetic mean is an index of precision (D). For example, if we can tolerate a standard error equal to 20% of the mean (a reasonable error in most bottom samples), then D is given by the general formula:

$$D = 0.2 = \frac{\text{standard error}}{\text{arithmetic mean}} = \frac{1}{\bar{x}}\sqrt{\frac{s^2}{n}}$$

Therefore the number of sampling units in a random sample is given by:

$$n = \frac{s^2}{D^2 \bar{x}^2} = \frac{s^2}{0.2^2 \bar{x}^2} = \frac{25s^2}{\bar{x}^2} \text{ for a 20\% error}$$

If the Poisson series is known to be a suitable model for the samples, then D is given by:

$$D = \frac{1}{\bar{x}}\sqrt{\frac{\bar{x}}{n}} = \frac{1}{\sqrt{n\bar{x}}}$$

Note that $n\bar{x} = \Sigma x =$ total count for the sample. Therefore the precision of the estimated population mean depends upon the total number ($n\bar{x}$) of animals in the sample, rather than sample size. For example, for a tolerable error of 20%:

$$D = 0.2 = \frac{1}{\sqrt{n\bar{x}}}$$

Therefore

$$n\bar{x} = \frac{1}{D} = \frac{1}{0.2^2} = 25$$

Therefore the product $n\bar{x}$ ($= \Sigma x$) must be always 25 for an error of 20%. The number of sampling units in a random sample is given by:

$$n = \frac{\bar{x}}{D^2 \bar{x}^2} = \frac{1}{D^2 \bar{x}} = \frac{25}{\bar{x}} \text{ for a 20\% error}$$

Therefore optimum sample sizes (to nearest integer) for a 20% error and various values of $\bar{x}$ are:

$\bar{x}$	1	5	10	20	25
n	25	5	3	2	1

If the negative binomial is known to be a suitable model for the samples, then D is given by:

$$D = \frac{1}{\bar{x}} \sqrt{\frac{\bar{x}}{n} + \frac{\bar{x}^2}{nk}} = \sqrt{\frac{1}{n\bar{x}} + \frac{1}{nk}}$$

and $\quad n = \frac{1}{D^2}\left(\frac{1}{\bar{x}} + \frac{1}{k}\right) = 25\left(\frac{1}{\bar{x}} + \frac{1}{k}\right)$ for a 20% error

When D is the relative error in terms of percentage confidence limits of the mean, the above formulae must be multiplied by t^2, where t is found in Student's t-distribution ($t \simeq 2$ for 95% probability level of D). For example, if we can tolerate 95% confidence limits of $\pm 40\%$ of the mean (equivalent to standard error of about 20% of the mean), then $D = 0.4$. Therefore the sample size needed to obtain an estimate of the population mean within $\pm 40\%$ of the true value is given by the general formula:

$$n = \frac{t^2 s^2}{D^2 \bar{x}^2} = \frac{2^2 s^2}{0.4^2 \bar{x}^2} = \frac{25 s^2}{\bar{x}^2}$$

If the Poisson series is a suitable model for the samples, then:

$$n = \frac{t^2}{D^2 \bar{x}} = \frac{2^2}{0.4^2 \bar{x}} = \frac{25}{\bar{x}}$$

If the negative binomial is a suitable model for the samples, then:

$$n = \frac{t^2}{D^2}\left(\frac{1}{\bar{x}} + \frac{1}{k}\right) = \frac{2^2}{0.4^2}\left(\frac{1}{\bar{x}} + \frac{1}{k}\right) = 25\left(\frac{1}{\bar{x}} + \frac{1}{k}\right)$$

These formulae require either $\bar{x}$ and s^2 (for general formula), $\bar{x}$ (for Poisson), or $\bar{x}$ and common k (for negative binomial). Values of these statistics are either guessed from the results of previous sampling, or calculated from a preliminary sample. As the value of n varies considerably throughout the year and with different species, a rough estimate of optimum sample size is adequate for most purposes. Therefore approximations assuming a normal distribution can be risked, and the general formula is suitable for most purposes. The special formulae for a Poisson series and negative binomial are easier to apply when either of these distributions is a suitable model for the samples. For example, if the negative binomial is a suitable model and a rough estimate of a common k is 2 (obtained from a preliminary sample), then the optimum sample sizes (to nearest integer) for 95% confidence limits of $\pm 40\%$ of the following means are:

$\bar{x}$	0·5	1	5	10	20	50	∞
n	63	38	18	15	14	13	13

If the tolerable 95% confidence limits are reduced to $\pm 20\%$ (equivalent to standard error of about 10% of the mean), then four times as many sampling units have to be taken, *e.g.* check that $n = 60$ for $\bar{x} = 10$ when $D = 0.2$ and $t = 2$. It is thus possible to calculate optimum sample sizes for various combinations of D and $\bar{x}$.

8.2.3 *The location of sampling units in the sampling area*

The terms *random sample* and *sampling units taken at random* are frequently used in this account. As many statistical methods require a random sample, it is important to understand the principles of random sampling.

The defined area for the investigation is the total available sampling area, and the number of available sampling units in this area is given by:

$$N = \frac{A}{a}$$

where A is the total sampling area and a is the area of the sampling unit (= quadrat size). Therefore N sampling units form the population from which a sample of n sampling units is selected. As n is usually much smaller than N, we must decide how to select the small sample of n units from the large population of N units. The sample must be representative of the whole population, and therefore the sampling units must be selected without bias. These conditions are fulfilled when the sampling units are selected at random from the population.

In simple random sampling, every sampling unit in the population has an equal chance of selection. True random selection is often difficult to achieve and the most reliable method is to use a table of random numbers. Suitable tables are included in many statistical textbooks and a large table is given in Fisher & Yates (1963, Table 33). To use these tables, simply select the first n numbers with values less than N. If all the available units in the population are listed from 1 to N, the n units in each sample are easily located. This method is often laborious in a large population. Therefore it is easier to use a large two-dimensional grid with each square equal to the area of a sampling unit. Some squares of the grid are redundant when the sampling area is irregular. The squares on two adjacent sides of the grid are numbered, and each sampling unit is thus located by a pair of co-ordinates. Random numbers are then drawn in pairs and these are used as co-ordinates to locate each unit in the sample. This is simply done by laying down two graduated lines at right angles, or by pacing out the required co-ordinates when the sampling area is large.

As nearly all the randomly-selected units in a sample may fall in one part of the sampling area, simple random sampling is not very efficient, especially when n is much smaller than N. Therefore the more efficient method of *stratified random sampling* is always preferable to simple random sampling. The purpose of stratified sampling is to increase sampling efficiency by dividing the population into several sub-populations or *strata*. These strata should be more homogeneous than the whole population, and should be well defined areas of known size. Stratification also increases the accuracy of population estimates and ensures that subdivisions of the population are adequately represented. The data from the different strata can be compared by a one-way analysis of variance (sources of variation: between strata, within strata).

In the simplest form of stratification, the whole sampling area is divided into areas of equal size (= strata). All the units in the sample are divided equally between strata and these units are located at random in each stratum. If the strata are unequal in area, the units in the sample are divided unequally between strata and the number of units allocated to each stratum is proportional to the area of the stratum, *i.e.* the total number of available units in the stratum. With proportional allocation of the units in a sample, the sampling fraction in each stratum is the same, *i.e.*

$$\frac{n_1}{N_1} = \frac{n_2}{N_2} = \cdots = \frac{n_k}{N_k} = \frac{n}{N}$$

where we select random sub-samples of $n_1, n_2, ..., n_k$ units from k strata containing $N_1, N_2, ..., N_k$ units. The total numbers of sampling units in the sample (n) and in the population (N) are given by:

$$n = n_1 + n_2 + \cdots + n_k$$
$$N = N_1 + N_2 + \cdots + N_k$$

Note that:

$$\frac{N_1}{N} = \frac{n_1}{n}, \quad \frac{N_2}{N} = \frac{n_2}{n}, \quad ..., \quad \frac{N_k}{N} = \frac{n_k}{n}$$

where $N_1/N, N_2/N, ..., N_k/N$ are the relative *weights* attached to each stratum. With proportional stratified sampling, the sample is *self-weighting* and the arithmetic mean of the whole sample is the best estimate of the population mean. This is calculated in the usual way, or from the means of the k strata:

$$\bar{x} = \frac{\Sigma x}{n} = \frac{n_1 \bar{x}_1 + n_2 \bar{x}_2 + \cdots + n_k \bar{x}_k}{n}$$

where $\bar{x}_1, \bar{x}_2, ..., \bar{x}_k$ are the arithmetic means of the different strata.

Standard error of mean = $(1/n)\sqrt{n_1 s_1^2 + n_2 s_2^2 + \cdots + n_k s_k^2}$
where $s_1^2, s_2^2, ..., s_k^2$ are the variances of the different strata. If the sampling fraction exceeds 10% ($n/N > 0.1$), then:

Standard error = $(1/n)\sqrt{(n_1 s_1^2 + n_2 s_2^2 + \cdots + n_k s_k^2)(1 - n/N)}$
where $(1 - n/N)$ is the *finite population correction*.

Although proportional allocation is most frequently used in stratified sampling, the theoretical optimum allocation of the units in the sample is the one that minimizes the standard error of the estimated mean for a given total cost of taking the sample. This is achieved when the sampling fraction for each stratum is proportional to the standard deviation for the stratum. As the standard deviations for the strata are rarely known before sampling, the method of optimum allocation is rarely used. With optimum allocation, the sample is no longer self-weighting, and therefore a weighted mean and standard error must be calculated (Snedecor & Cochran 1967).

Stratified sampling is of greatest value when the sampling area contains a diversity of biotopes. For example, we wish to sample a section of river with a total sampling area of 200 m², and the area of the sampling unit is 0.05 m². Therefore 4000 sampling units form the population ($N = 4000$) from which a random sample of 40 units must be selected ($n = 40$). The following five strata ($k = 5$) were recognised (units in each stratum given in parentheses): Large stones on gravel ($N_1 = 1000$), Gravel ($N_2 = 500$), Plant A on gravel ($N_3 = 1500$),

Plant B on gravel ($N_4 = 800$), Mud ($N_5 = 200$). As the standard deviations for the strata are not known, optimum allocation of units cannot be used. Therefore the units in the random sample ($n = 40$) are allocated in proportion to the areas of the strata thus:

$$n_1 = 10 \quad n_2 = 5 \quad n_3 = 15 \quad n_4 = 8 \quad n_5 = 2$$

These units are selected at random from the available sampling units in each stratum. Note that the sampling fraction is the same for each stratum ($n/N = 0.01$).

Other methods for selecting a sample from a population have a more limited application than random sampling. If the object of the investigation is to determine the mean and variance of the population, random sampling is essential. If the object is to determine numbers in relation to position within the ecosystem, *systematic sampling* may be preferable. In systematic sampling, the first unit in the sample is selected at random from the population and then the next units are selected at fixed intervals, *e.g.* every 10th unit in the population is taken until the required sample size is obtained. The advantages of systematic sampling are:

(1) it is easy to draw a sample, since only one random number is required;
(2) the units in the sample are distributed evenly throughout the population.

The disadvantages are:

(1) the sample may be very biased when the interval between units in the sample coincides with a periodic variation in the population;
(2) there is no reliable method for estimating the standard error of the sample mean.

A variation of systematic sampling is the *centric systematic area-sample* which can be treated as a random sample (Milne 1959). In this method, each sampling unit is taken from the exact centre of each stratum.

Another method is to use a grid of contiguous quadrats (Greig-Smith 1964). This method is used in plant ecology and facilitates the detection of mosaic patterns in non-random distributions of species. The method is limited to small areas of bottom and to biotopes which are suitable for a contiguous arrangement of quadrats. Another method of mapping aggregations is to take sampling units in pairs with one unit located at random and the second unit located at a fixed distance from the first (Hughes 1962).

8.3 SUB-SAMPLING IN THE LABORATORY WITH LARGE CATCHES

Although it is usually possible to count all the invertebrates in a sampling unit, the task is often laborious when the catch per sampling unit is very large. This problem can be solved by reducing the size of the sampling unit, but this is often impossible. An alternative solution is *sub-sampling* (also called *two-stage sampling*). First, a sample of n_1 *primary sampling units* is selected in the usual way from the population in the field (see section 8.2.3). In the second stage, a sub-sample of n_2 *sub-units* (or *second-stage units*) is taken from the total catch for each primary unit. The large catch for each primary unit is concentrated in a known volume of water (or preservative), and is well agitated in a container before sub-sampling. If the invertebrates are distributed randomly in the container before sub-sampling and only a small proportion of the total catch is removed in each sub-unit, then the counts of the sub-sample should be distributed according to a Poisson series. This hypothesis should be checked by a χ^2 test (variance to mean ratio, section 4.1.2) on a sub-sample of at least five sub-units. If agreement with a Poisson series is accepted, then it can be assumed that the invertebrates are distributed randomly before sub-sampling, and subsequently a sub-sample of only one sub-unit need be taken. Thus a single count can be used to estimate the total numbers for each primary sampling unit, and the accuracy of this estimate depends upon the size of the count (section 6.2.2).

Sub-sampling is used extensively in the quantitative study of plankton Populations and several papers provide a detailed account of statistical methods (*e.g.* Ricker 1937, Holmes & Widrig 1956, Kutkuhn 1958, Lund, Kipling & Le Cren 1958). The following imaginary example outlines the chief statistical methods.

Large numbers of chironomid larvae were taken in a random sample of 10 (primary) sampling units ($n_1 = 10$). The other less-numerous invertebrates were removed from the total catch for each primary unit, and then chironomid larvae plus detritus were transferred to 4 litres of preservative in a flask. This process was repeated until there were 10 flasks, each containing the total catch of chironomid larvae for one primary unit. After thorough agitation of a flask, a sub-unit of 50 ml was removed with a pipette. This process was repeated until a sub-sample of 5 sub-units had been taken from one flask. Chironomid larvae in each sub-unit were counted, and the following counts were obtained for the first sub-sample:

$$20, 25, 25, 30, 40; \quad \bar{x} = 28, \quad s^2 = 57 \cdot 5, \quad n_2 = 5.$$

The χ^2 test (variance to mean ratio, section 4.1.2) gave a value of $\chi^2 = 8 \cdot 2$ with 4 degrees of freedom. As this χ^2 value clearly lies between the 5% significance levels in Fig. 8, agreement with a Poisson series is accepted and it is assumed that the chironomid larvae were distributed randomly in the flask before sub-sampling. Therefore we can now estimate the total numbers of larvae in the first flask, *i.e.* total numbers for the first primary sampling unit. The accuracy of this estimate increases as the size of the count increases. Therefore the total count for the sub-sample (*i.e.* 140) is used and 95% confidence limits for this count are 117 to 163 (from Crow & Gardner 1959). As the total count is for a volume of 250 ml (5 sub-units of 50 ml each) or $\frac{1}{16}$ of the total volume in the flask (4 litres), the count and its confidence limits are multiplied by 16 to give an estimate of the total numbers in the first flask. Therefore the estimated number of chironomid larvae for the first primary unit is 2240 with 95% confidence limits of 1872 and 2608 (or 2240 ± 368).

These calculations are now repeated for each flask and hence for each primary sampling unit. Sufficient sub-units should be taken to ensure that the total count for each sub-sample is at least 100 (for 95% confidence limits of $\pm 20\%$; see section 6.2.2). As the estimation of total numbers and confidence limits by the above methods requires agreement with a Poisson series, it is important to ensure that the invertebrates are distributed randomly before sub-sampling. Therefore various methods of mixing should be tried on preliminary samples until a satisfactory technique is found.

If the counts of a sub-sample do not follow a Poisson series and the variance of the sub-sample is significantly greater than the mean, then the invertebrates were distributed contagiously before sub-sampling. It is now more difficult to estimate total numbers and confidence limits for each primary sampling unit. The simplest method is to transform the counts of a sub-sample to logarithms, and then calculate the geometric mean with its 95% confidence limits (see section 6.2.4, example 22).

8.4 SUMMARY GUIDE

Define the objects of the study and the area in which the study takes place. Most studies fall into two broad categories:

(A) *Faunal surveys* (section 8.1): Chief objects are to discover which species are present, and to estimate the relative abundance of each species at different stations in the sampling area. Therefore the sample at each station should cover a large area of bottom.

(B) *Quantitative studies* (section 8.2): Chief object is to estimate numbers per unit area for each species, and therefore quantitative comparisons can be made. Major considerations are:

(1) The dimensions of the sampling unit (quadrat size). The smallest possible sampling unit should be used, but the chosen quadrat size is always a compromise between statistical and practical requirements (section 8.2.1).

(2) The number of sampling units in each sample. Large samples ($n > 50$) are preferable, but the size of small samples can be calculated for a specified degree of precision:

$$n = \frac{s^2}{D^2 \bar{x}^2}$$

where D is an index of precision, *i.e.* the required standard error as a proportion of the mean. This formula is multiplied by t^2 ($t \simeq 2$) when D is the relative error in terms of percentage confidence limits of the mean. Approximate values of $\bar{x}$ and s^2 are guessed from previous samples or a preliminary sample. There are special formulae for a Poisson series and negative binomial when either of these distributions is a suitable model for the samples (section 8.2.2).

(3) The location of sampling units in the sampling area. The units in the sample are usually located at random in the sampling area and all the available units in the population must have an equal chance of selection for the sample. A table of random numbers is used to select the units for the sample. The sample must be representative of the whole population, and therefore stratified random sampling is preferable to simple random sampling. The sampling area (= population) is divided up into several strata (= subpopulations). These strata can be unequal in area, *e.g.* when the strata are different biotopes. The units in the sample are allocated to each stratum in proportion to the area of the stratum. The arithmetic mean of the whole sample is the best estimate of the population mean.

Other methods for selecting a sample from a population are briefly reviewed (section 8.2.3), and the method of sub-sampling in the laboratory is described (section 8.3).

IX ACKNOWLEDGMENTS

I wish to express my sincere thanks to the following for their help in the preparation of this booklet: Mr. H. C. Gilson for his constant help and encouragement; Mr. J. N. R. Jeffers, F.I.S., who has read the whole of this booklet in manuscript and made many suggestions which have greatly improved the final text; T. B. Bagenal, Professor R. W. Edwards, Dr. J. G. Jones, Charlotte Kipling, Dr. T. T. Macan, E. D. LeCren, Dr. D. W. Sutcliffe for reading all or part of the manuscript; Dr. T. T. Macan for supplying the original data for example 22B; A. E. Ramsbottom for drawing all the figures; Mrs. D. Parr and Mrs. P. A. Tullett for all their assistance in this work. If I have inadvertently adopted or adapted where I should have sought permission, I hope it will be excused as oversight or ignorance.

REFERENCES

Albrecht, M. L. (1959). Die quantitive Untersuchung der Bodenfauna fließender Gewässer. *Z. Fisch.* **8**, 481-550.

Anscombe, F. J. (1949). The statistical analysis of insect counts based on the negative binomial distribution. *Biometrics*, **5**, 165-173.

Anscombe, F. J. (1950). Sampling theory of the negative binomial and logarithmic series distributions. *Biometrika*, **37**, 358-382.

Arbous, A. G. & Kerrich, J. E. (1951). Accident statistics and the concept of accident-proneness. *Biometrics*, **7**, 340-432.

Bagenal, M. (1955). A note on the relations of certain parameters following a logarithmic transformation. *J. mar. biol. Ass. U.K.* **34**, 289-296.

Bailey, N. T. J. (1959). *Statistical Methods in Biology.* London.

Beall, G. (1939). Methods of estimating the population of insects in a field. *Biometrika*, **30**, 422-439.

Beall, G. (1940). The fit and significance of contagious distributions when applied to observations on larval insects. *Ecology*, **21**, 460-474.

Bishop, O. N. (1966). *Statistics for Biology.* London.

Bliss, C. I. & Fisher, R. A. (1953). Fitting the binomial distribution to biological data and a note on the efficient fitting of the negative binomial. *Biometrics*, **9**, 176-200.

Bliss, C. I. & Owen, A. R. G. (1958). Negative binomial distributions with a common k. *Biometrika*, **45**, 37-58.

Campbell, R. C. (1967). *Statistics for Biologists.* Cambridge.

Cassie, R. M. (1962). Frequency distribution models in the ecology of plankton and other organisms. *J. Anim. Ecol.* **31**, 65-92.

Cochran, W. G. (1954). Some methods for strengthening the common χ^2 tests. *Biometrics*, **10**, 417-451.

Cochran, W. G. (1963). *Sampling Techniques.* (2nd edition). New York.

Cole, L. C. (1946). A theory for analyzing contagiously distributed populations. *Ecology*, **27**, 329-341.

Colebrook, J. M. (1960). Plankton and water movements in Windermere. *J. Anim. Ecol.* **29**, 217-240.

Comita, G. W. & Comita, J. J. (1957). The internal distribution patterns of a calanoid copepod population, and a description of a modified Clarke-Bumpus plankton sampler. *Limnol. Oceanogr.* **2**, 321-332.

Comrie, L. J. (1948-49). Chambers's Six-figure Mathematical Tables. Vols. 1 & 2, Edinburgh.

Crow, E. L. & Gardner, R. S. (1959). Table of confidence limits for the expectation of a Poisson variable. *Biometrika*, **46**, 441-453. (New Statistical Tables Series No. 28.)

Cummins, K. W. (1962). An evaluation of some techniques for the collection and analysis of benthic samples with special emphasis on lotic waters. *Am. Midl. Nat.* **67**, 477-504.

REFERENCES

David, F. N. (1953). *A Statistical Primer.* London.

David, F. N. & Moore, P. G. (1954). Notes on contagious distributions in plant populations. *Ann. Bot.* **18**, 47–53.

Davies, H. T. (1935). *Tables of the Higher Mathematical Functions* (2). Bloomington, Ind.

Debauche, H. R. (1962). The structural analysis of animal communities in the soil. In *Progress in Soil Zoology* (ed. P. W. Murphy), pp. 10–25. London.

Documenta Geigy (1962). *Scientific Tables.* 6th ed. (ed. K. Diem). Manchester.

Edgar, W. D. & Meadows, P. S. (1969). Case construction, movement, spatial distribution and substrata selection in the larva of *Chironomus riparius* Meigen. *J. exp. Biol.* **50**, 247–253.

Evans, D. A. (1953). Experimental evidence concerning contagious distributions in ecology. *Biometrika*, **40**, 186–211.

Finney, D. J. (1946). Field sampling for the estimation of wireworm populations. *Biometrics*, **2**, 1, 1–7.

Fisher, R. A., Corbet, A. S. & Williams, C. B. (1943). The relation between the number of species and the number of individuals in a random sample of an animal population. *J. Anim. Ecol.* **12**, 42–58.

Fisher, R. A. & Yates, F. (1963). *Statistical Tables for Biological, Agricultural and Medical Research.* (6th ed.) Edinburgh.

Gaufin, A. R., Harris, E. K. & Walter, H. J. (1956). A statistical evaluation of bottom sampling data obtained from three standard samplers. *Ecology*, **37**, 643–648.

Green, R. H. (1966). Measurement of non-randomness in spatial distributions. *Researches Popul. Ecol. Kyoto Univ.* (*I*), **8**, 1–7.

Greig-Smith, P. (1964). *Quantitative Plant Ecology.* (2nd ed.) London.

Harris, E. K. (1957). Further results in the statistical analysis of stream sampling. *Ecology*, **38**, 463–468.

Hartenstein, R. (1961). On the distribution of forest soil microarthropods and their fit to "contagious" distribution functions. *Ecology*, **42**, 190–194.

Healy, M. J. R. & Taylor, L. R. (1962). Tables for power-law transformations. *Biometrika*, **49**, 557–9.

Heywood, J. & Edwards, R. W. (1961). Some aspects of the ecology of *Potamopyrgus jenkinsi* Smith. *J. Anim. Ecol.* **31**, 239–250.

Holmes, R. W. & Widrig, T. M. (1956). The enumeration and collection of marine phytoplankton. *J. Cons. perm. int. Explor. Mer*, **22**, 21–32.

Hughes, R. D. (1962). The study of aggregated populations. In *Progress in soil zoology* (ed. P. W. Murphy), pp. 51–55. London.

Iwao, S. & Kuno, E. (1968). Use of the regression of mean crowding on mean density for estimating sample size and the transformation of data for the analysis of variance. *Researches Popul. Ecol. Kyoto Univ.* **10**, 210–214.

Iwao, S. (1968). A new regression method for analyzing the aggregation pattern of animal populations. *Researches Popul. Ecol. Kyoto Univ.* **10**, 1–20.

Jones, P. C. T., Mollison, J. E. & Quenouille, M. E. (1948). A technique for the quantitative estimation of soil micro-organisms. *J. gen. Microbiol.* **2**, 54–69.

REFERENCES

Kendall, M. G. (1962). *Rank Correlation Methods.* London.
Kutkuhn, J. H. (1958). Notes on the precision of numerical and volumetric plankton estimates from small-sample concentrates. *Limnol. Oceanogr.* **3**, 69-83.
Lefkovitch, L. P. (1966). An index of spatial distribution. *Researches Popul. Ecol. Kyoto Univ.* **8**, 89-92.
Lewis, T. & Taylor, L. R. (1967). *Introduction to Experimental Ecology.* London and New York.
Lloyd, M. (1967). "Mean crowding". *J. Anim. Ecol.* **36**, 1-30.
Lund, J. W. G., Kipling, C. & Le Cren, E. D. (1958). The inverted microscope method of estimating algal numbers and the statistical basis of estimations by counting. *Hydrobiologia*, **11**, 143-170.
Macan, T. T. (1958). Methods of sampling the bottom fauna in stony streams. *Mitt. int. Verein. theor. angew. Limnol.* **8**, 1-21.
Macan, T. T. (1959). *A Guide to Freshwater Invertebrate Animals.* London.
Mainland, D., Herrera, L. & Sutcliffe, M. I. (1956). *Statistical Tables for Use with Binomial Samples—Contingency Tests, Confidence Limits and Sample Size Estimates.* New York.
Milne, A. (1959). The centric systematic area-sample treated as a random sample. *Biometrics*, **15**, 270-297.
Morisita, M. (1959). Measuring the dispersion of individuals and analysis of the distributional patterns. *Mem. Fac. Sci. Kyushu Univ. Ser. E. Biol.* **2**, 215-235.
Moroney, M. J. (1956). *Facts from Figures.* Harmondsworth (3rd ed.).
National Bureau of Standards (1950). *Tables of the Binomial Probability Distribution. Applied Maths.* series no. 6. Washington.
Neyman, J. (1939). On a new class of "contagious" distributions, applicable in entomology and bacteriology. *Ann. math. Statist.* **10**, 35-57.
Pearson, E. S. & Hartley, H. O. (1966). *Biometrika Tables for Statisticians.* (3rd ed.) Cambridge.
Pólya, G. (1931). Sur quelques points de la théorie des probabilités. *Annls Inst. Henri Poincaré*, **1**, 117-161.
Quenouille, M. H. (1949). A relation between the logarithmic, Poisson, and negative binomial series. *Biometrics*, **5**, 162-4.
Quenouille, M. H. (1950). *Introductory Statistics.* London.
Quenouille, M. H. (1959). *Rapid Statistical Calculations.* London.
Ricker, W. E. (1937). Statistical treatment of sampling processes useful in the enumeration of plankton organisms. *Arch. Hydrobiol.* **31**, 68-84.
Siegel, S. (1956). *Non-parametric Statistics for the Behavioral Sciences.* New York.
Skellam, J. G. (1952). Studies in statistical ecology. 1. Spatial pattern. *Biometrika*, **39**, 346-362.
Snedecor, G. W. & Cochran, W. G. (1967). *Statistical Methods.* Ames, Iowa.
Southwood, T. R. E. (1966). *Ecological Methods.* London.
Stuart, A. (1962). *Basic Ideas of Scientific Sampling.* London.

Taylor, C. C. (1953). Nature of variability in trawl catches. *Fishery Bull. Fish Wildl. Serv. U.S.* **54**, 145–166.

Taylor, L. R. (1961). Aggregation, variance and the mean. *Nature, Lond.* **189**, 732–5.

Thomas, M. (1949). A generalization of Poisson's binomial limit for use in ecology. *Biometrika*, **36**, 18–25.

Thomson, G. W. (1952). Measures of plant aggregation based on contagious distribution. *Contr. Lab. vertebr. Biol. Univ. Mich.* **53**, 1–16.

Welch, P. S. (1948). *Limnological Methods.* Philadelphia.

Wilcoxon, F. & Wilcox, R. A. (1964). *Some Rapid Approximate Statistical Procedures.* New York.

Williams, C. B. (1964). *Patterns in the Balance of Nature.* London & New York.

Williamson, E. & Bretherton, M. H. (1963). *Tables of the Negative Binomial Probability Distribution.* New York.

Yates, F. (1960). *Sampling Methods for Censuses and Surveys.* (3rd ed.). London.

APPENDIX: SYMBOLS AND TERMS

1. Mathematical symbols

Symbol	Meaning
$>$	greater than
$<$	less than
$\simeq$	is approximately equal to
$\geq$	is greater than or equal to
$\leq$	is less than or equal to
$\pm$	plus or minus
$\rightarrow$	tends to
$\vert\ \vert$	absolute value of term between vertical lines
$!$	factorial; $4! = 4 \times 3 \times 2 \times 1$
$\wedge$	"hat"; indicates an estimate of a term
∞	infinity
$\sqrt{\ }$	square root of

2. Greek symbols

Symbol	Meaning
α	alpha
β	beta, angle of elevation for regression line
δ	delta
λ	lambda; Poisson parameter
μ	mu; arithmetic mean of population
Σ	sigma; sum of
σ	sigma; standard deviation
σ^2	sigma squared; variance
χ^2	chi squared

3. Latin symbols

Symbol	Meaning
A	total sampling area
a	area of sampling unit (equals quadrat size)
$A_{(x)}$	total number of counts exceeding x in negative binomial
a	parameter in Taylor's power law
b	regression coefficient and exponent in Taylor's power law
C	coefficient of variation
c	either the statistic m or a single count from a Poisson series
D	index of precision (ratio of standard error to arithmetic mean)
d	standardised deviate or normal variable with zero mean and unit standard deviation
e, $\log_e$ ln	base of natural (Napierian) logarithms
F	variance ratio (F-test)
f	frequency
H_0	null hypothesis
I	index of dispersion

APPENDIX

I_δ	Morisita's index of dispersion
K	test statistic (Kruskal–Wallis one-way analysis by ranks)
k	index of clumping in a population; exponent in positive and negative binomials
k	number of groups in a χ^2 contingency table
k_c	common k in negative binomial
$\hat{k}$	estimate of k
M	test criterion in Bartlett's test
m	Poisson statistic ($=\bar{x}=s^2$)
$\overset{*}{m}$	index of "mean crowding"
N	total number of available sampling units in sampling area
n	total number of sampling units in sample
P	probability or level of significance
	$* = P<0.05$ Significant
	$** = P<0.01$ Highly significant
	$*** = P<0.001$ Very highly significant
p	probability of a positive response in a binomial distribution; index in transformation of Taylor's power law
q	probability of a negative response in a binomial distribution $q = 1 - p$
R	sum of ranks
r	product-moment correlation coefficient
r_s	Spearman's rank correlation coefficient
S	test statistic (Friedman's two-way analysis by ranks)
s	standard deviation of sample
s^2	variance of sample
T	difference between observed and expected 3rd moment in negative binomial
t	Student's t-statistic
U	difference between observed and expected variance in negative binomial
U	test statistic in Mann–Whitney U-test
v	number of degrees of freedom
x	variate or independent variable in regression analysis (also called regressor variable)
$\bar{x}$	arithmetic mean of sample
$\overset{*}{x}$	estimate of $\overset{*}{m}$
y	transformed variate or dependent variable in regression analysis
$\bar{y}$	arithmetic mean of transformed counts.

INDEX

Analysis of variance, 96
 explanation of, 98
 application to small samples, 107
Arithmetic mean,
 definition of, 11
 calculation of, 12

b of Taylor's Power Law, 76
Binomial family,
 relationships between members of, 36
 over-dispersion — see contagious distribution
 under-dispersion — see regular distribution
 uniform distribution — see regular distribution
 even distribution — see regular distribution
 clumped distribution — see contagious distribution
 aggregated distribution — see contagious distribution

Central-limit theorem, 95, 105
Centric systematic area-sample, 134
Charlier coefficient, 74
chi-squared (χ^2), 40, 44, 121
Class frequency, 14
Clumping—see Negative binomial, 23
Coefficient of variation, 98
Cole's Index, 77
Comparison of samples (parametric tests), 94
 large samples, 95
 comparison of means—large samples, 96
 comparison of variances—large samples, 97
 comparison of means—Poisson series, 103

comparison of means — small samples from contagious distributions (t-tests, F-tests and analysis of variance), 105
 analysis of variance, 96
 coefficient of variation, 98
 correlation coefficient, 102
 regression analysis, 63, 72, 102
 calculations for linear regression, 73
Comparison of samples (non-parametric tests),
 explanation of, 112
 comparison of two samples (Mann-Whitney U-test), 113
 comparison of more than two samples (one-way classification), 115
 comparison of more than two samples (two-way classification), 118
 correlation coefficient, 120
 χ^2 and contingency tables, 121
Comparison of samples, summary of, 125
Compound Poisson distribution, 52
Confidence limits of mean,
 definition of, 81
 calculation for large samples, 81
 calculation for small sample from Poisson, 83
 calculation for small sample from positive binomial, 86
 calculation for small sample from a contagious distribution, 86
 calculation for samples from negative binomial, 87
 calculation for samples with Taylor's Power Law, 88
 calculation for samples with log transformation, 90
 summary, 92

INDEX

Contagious distribution, 37
  definition of, 50
  mathematical models for confidence limits (log transformation), 90
  comparison of small samples (transformations), 105
  see also Negative binomial
Continuous variable
  definition of, 14
Correlation coefficient, 102, 120

David and Moore's Index, 74
Degrees of freedom, 13
Derived mean, 33
Deviation—see Variance
Discontinuous variable,
  definition of, 14
Discrete log-normal, 66
Distribution-free, see Comparison of samples (non-parametric tests)

Examples, list of, 6

F-test, 105
Faunal surveys, 126
Frequency classes, 14
Frequency distribution,
  arrangements of counts in, 14
  calculation of mean and variants from, 14
Friedman's test, 118

Geometric mean, 12
Green's coefficient, 74

Heterogeneous Poisson sampling, 52
Histogram, definition of, 14

Indices of dispersion, explanation of, 73
  b of Taylor's Power Law, 76
  Charlier coefficient, 74
  Cole's Index, 77
  David and Moore's Index, 74
  Green's coefficient, 74
  Index of Lexis, 74
  k in negative binomial, 75
  Lefkovitch's Index, 77
  Lloyd's index of mean crowding, 78
  Morisita's Index, 76
  variance to mean ratio, 40, 74

k in negative binomial, 75
Kendall's rank correlation coefficient, 121
Kruskal-Wallis one-way analysis by ranks, 117

Lefkovitch's Index, 77
Lexis, index of, 74
Linear regression, calculation of, 73
Lloyd's index of mean crowding, 78
Logarithmic graph paper, use of, 71
Logarithmic series, 23
Logarithmic transformation, uses for:
  contagious distribution, 32
  confidence limits, 90
  analysis of variance, 108

Mann-Whitney U-test, 113
Maximum-likelihood, method of, 24
Median, 12
Moments, 54
Morisita's index of dispersion, 70, 76

Natural (Napierian) logarithms, 24
Negative binomial, 16
  definition of, 23
  calculation of constant k and expected frequencies (large samples), 25

INDEX

use as mathematical model, 51
test for agreement with (large samples), 53
test for agreement with (small samples), 54
estimation of constant k (small samples), 55
estimation of common k, 63
use of k as index of dispersion, 75
confidence limits of mean, 87
Neyman Type A distribution, 66
Non-parametric tests, see Comparison of samples
Normal distribution, definition of, 30
Null-hypothesis, 94

Occam's razor, 40

Parameter, definition of, 13
Parametric tests, see Comparison of samples
Pearson Type III distribution, 52
Poisson series
  definition of, 18
  calculation of expected frequencies, 19
  conditions for, 22
  rapid test for agreement with, 40
  variance to mean ratio test, 40
  χ^2 goodness of fit test, 44
  confidence limits of mean, 83
  comparison of means, 103
Polya-Aeppli distribution, 66
Population (in ecology), 11
Population (in statistical terminology), 11
Positive binomial,
  definition of, 17
  test for agreement with, 47
  calculation of expected frequencies, 18
  calculation of confidence limits, 86
Probability, definition of, 17

Quadrat size, effect of, 39, 68
Quantitative sampling, 127
  quadrat size, 39, 128
  number of sampling units, 128
  random sampling, 131
  stratified random sampling, 132
  systematic sampling, 134
  centric systematic area-sample, 134
  sub-sampling, 135
  summary to sampling, 136
Quenouille's test, 115

Random distribution, 37
  definition of, 38
  rapid test, 40
  variance to mean ratio test, 40
  χ^2 goodness of fit test, 44
  summary of tests, 48
  confidence limits of mean, 83
  comparison of means, 103
Random sampling, 131
Rank order, see Comparison of samples
Regression analysis, 63, 72, 102
Regular distribution, 37
  definition of, 46
  test for agreement (Positive binomial), 46
  calculation of confidence limits, 86

Sample, definition of, 11
Sampling programme, 126
  faunal surveys, 126
  quantitative studies, 127
Sampling unit, definition of, 11
Significance, levels of, 94
Skewness, 66
Spatial dispersion, 37
Spearman's rank-correlation coefficient, 121
Standard deviation, 13
Standard error, 80
Statistic, definition of, 13
Stratified random sample, 132

Sub-sampling, 135
Symbols and terms, 143
Systematic sampling, 134

t-distribution, use for confidence limits, 82
  use for comparison of means, 106
Taylor's Power Law,
  definition of, 71
  application of, 72
  use as index of dispersion, 76
  use for transformations, 71, 88, 106
Thomas's contagious distribution, 66

Transformations,
  reason for, 30
  selection of, 33
  effect of, 34
  logarithmic transformation, calculation of confidence limits, 90
  transformation in Taylor's Power Law, 88
  use in t-tests, F-tests, analysis of variance, 106
  checking adequacy of, 109

Variance, definition of, 12
  calculation of, 13